The Globalization of Clean Energy Technology

Urban and Industrial Environments

Series editor: Robert Gottlieb, Henry R. Luce Professor of Urban and Environmental Policy, Occidental College

For a complete list of books published in this series, please see the back of the book.

The Globalization of Clean Energy Technology
Lessons from China

Kelly Sims Gallagher

The MIT Press
Cambridge, Massachusetts
London, England

First MIT Press paperback edition, 2017

MIT Press books may be purchased at special quantity discounts for business or sales promotional use. For information, please email special_sales@mitpress.mit.edu.

This book was set in Sabon by Toppan Best-set Premedia Limited, Hong Kong. Printed and bound in the United States of America.

Library of Congress Cataloging-in-Publication Data

Gallagher, Kelly Sims.
The globalization of clean energy technology : lessons from China / Kelly Sims Gallagher.
    pages   cm. — (Urban and industrial environments)
Includes bibliographical references and index.
ISBN 978-0-262-02698-7 (hbk. : alk. paper)
ISBN 978-0-262-53373-7 (pb.)
1. Clean energy investment—China—Case studies. 2. Renewable energy sources—China—Case studies. 3. Renewable energy sources—Economic aspects—Case studies. 4. Business enterprises, Foreign—China—Case studies. 5. Energy policy—Economic aspects—Case studies. 6. Technology transfer—Case studies. I. Title.
HD9502.5.C543C6237   2014
333.79'40951—dc23
2013030857

10 9 8 7 6 5 4 3 2

For Kevin, Theo, and Estelle Gallagher

Contents

Acknowledgments

Since 2001, I've been engaged in empirical research on international technology transfer in cleaner energy technologies, and this book continues that work and draws out broader implications for theory and policy. The origins of this book grew out of conversations with many faculty and research fellows in the Energy Technology Innovation Policy (ETIP) research program in the Belfer Center for Science and International Affairs at the Harvard Kennedy School, which I directed from 2003 to 2009. An important mentor, John P. Holdren, founded the ETIP program, and he gave me invaluable advice and constructive criticism during that period. I want to thank Bill Clark for prodding me to think about the larger implications of my research, which gave me the germ of the idea for this book. I benefited from conversations with many colleagues there, especially Graham Allison, Laura Diaz Anadon, Matt Bunn, Calestous Juma, Henry Lee, and Ambuj Sagar.

I started to work on the book seriously when I moved to the Fletcher School in late 2009. Colleagues, research fellows, and students there provided encouragements, constructive criticism, and new ideas. I particularly thank Jenny Aker, Rishi Bhandary, Dan Drezner, Liu Hengwei, Laura Kuhl, Bill Moomaw, Kartikeya Singh, Peter Uvin, Xuan Xiaowei, and Zhang Fang. Numerous Fletcher students provided research assistance at various times, most notably Paolo Cozzi, Nick Davidson, Kiel Downey, Lasse Eisgruber, Erin Kempster, Prashanth Parameswaran, Ben Rabe, Michael Reading, J. R. Siegel, and Aaron Strong. Miranda Fasulo, Celia Mokalled, and Mieke van der Wansem brought a semblance of order to my work life, and enabled me to find blocks of time to focus on this project.

I served as a lead author of the Global Energy Assessment's chapter on innovation concurrently with the research and writing of this book.

My coauthors and I thoroughly reviewed and assessed all the literature on energy innovation during this period, and I benefited enormously from our discussions, many of which directly influenced this book.

Much of the new empirical contribution for this book came from research conducted in China, Germany, and the United States between 2010 and 2012. I particularly thank Professor Su Jun, dean of social sciences at Tsinghua University, and Professor Xue Lan, dean of the School of Public Policy and Management at Tsinghua, for welcoming me to serve as a visiting professor at Tsinghua University in 2010. They were not only superb hosts but also generous teachers, and warm friends to my family. Professor Su's research group was exemplary in every way. I especially thank Zhang Fang and Zhi Qiang for their research assistance. Ru Peng and Xia Di helped me navigate the logistics, and made our lives easier through dozens of thoughtful arrangements and creative ideas. Particular thanks to Zheng Fangneng at the China's Ministry of Science and Technology who helped arrange many interviews and factory visits. One of my earliest and best teachers about China, Alan Wachman, tragically passed away while I was writing this book. He not only taught me much about Chinese history but also how to conduct scholarship with care and humility. He was an unflagging cheerleader who kept writing notes of encouragement.

In Germany, I served as the 2012 EADS distinguished visitor at the American Academy of Berlin. The academy was immensely helpful in arranging meetings and interviews, and provided a delightful atmosphere in which to try out my ideas during my residency. I thank Ulrike Graalfs, Pamela Rosenberg, and Gary Smith especially for their efforts on my behalf.

All of the people who took time out of their busy schedules to talk with me, show me their facilities, and respond to my endless questions made this book possible.

Most of the book was written as a visiting scholar at the American Academy of Arts and Sciences, and I thank the Academy, especially John Randell, for providing me with an extraordinarily peaceful and pleasant place to think, read, and write. I also thank Bob Fri for bringing me into the orbit of the American Academy through his leadership of the Alternative Energy Future project.

I received wonderfully helpful comments from three anonymous reviewers. In addition, Kevin Gallagher, Bill Moomaw, and Zhang Fang also read complete drafts as well and made insightful suggestions that greatly improved the book. All remaining errors are my own. I warmly thank Clay Morgan at the MIT Press for his support throughout the process of writing this book. He provided vital encouragement at the outset, and shepherded the book through the publication process with skill and kindness.

I most gratefully acknowledge financial support, which was provided through grants from BP and the William and Flora Hewlett Foundation, a gift from Barbara Kates-Garnick and Marc B. Garnick, MD, and a teaching leave from the Fletcher School in spring 2012.

Three extraordinary friends helped to raise my kids during this period while also providing cheer and warmth—Carey and Peter Thomson and Alexandra Vacroux. Norma Gallagher, Jo Ann and Jim Gillula, Bill and Jennifer Sims, and Colin Sims similarly helped, and were unfailingly confident that I would get this done.

Most of all, I wish to thank my husband and children for their adventuresome spirit, curiosity, willingness to live in China while I was conducting my research there, and toleration of my extended research trips away from home. Throughout the process, Kevin alternately inspired and compelled me, and protected my time and space. My children, Theo and Estelle, were consistent sources of joy and laughter. They were remarkably tolerant of their busy and distracted mother, and always reminded me why I was writing this book.

1

Introduction

On a hot summer day in 2010, I ventured into the factory of a leading solar photovoltaic (PV) manufacturer in China. As I toured the facility, I noticed manufacturing equipment from many different countries installed as part of the assembly line. I registered with some dismay, however, that I did not see any equipment from my own country, the United States. Meanwhile, I was surprised to see how few people were working in the factory; aside from the ingot production area, the factory was sparkling clean, white, sparse, and almost fully automated. Where was the so-called labor cost advantage in China? After an exhaustive tour of the entire production line, my host, the chief technology officer (CTO) of this company, then invited me to lunch in the company cafeteria. With my back to the room, I slid down the buffet line, choosing between noodles and rice, various meat and vegetable dishes, and a selection of beverages. Suddenly, I realized that the foreign language being spoken behind me was not Chinese but German! I turned around and discovered dozens of German men in coveralls clustered around the tables taking their lunch break. "Who *are* they?" I asked. The CTO offhandedly replied, "Just technicians, helping to install equipment in our new assembly line in the new factory."

It was at that moment in my research when it became clear to me that the development and deployment of cleaner energy technologies is now truly a globalized phenomenon. Here was a Chinese CTO with a PhD from an Australian university supervising the assembly of a new factory line that was being constructed mostly with Japanese and German manufacturing equipment. German technicians were assembling parts of the line, which would be using core technology licensed from numerous countries around the world. The PV modules being packed by Chinese workers at the end of the assembly line were being shipped to countries all around the world, but especially to Germany, Spain, and the United

States, where local installation companies would mount them on homes, office buildings, and schools. These countries had all created large markets for solar PV technologies through market-formation policies.

To address the myriad challenges associated with energy consumption, ranging from air pollution and global climate disruption to energy insecurity and volatile oil prices, the accelerated and ubiquitous global diffusion of cleaner and more efficient energy technologies is urgently needed. Indeed, a basic assumption of this book is that the sooner and faster that such technologies are adopted and used around the world, the better.

But in policy circles, academia, and the business community, there is considerable debate about the barriers and incentives to the movement of these cleaner technologies across international borders. The debates are about which barriers are most daunting and difficult to bypass, which are created by governments, firms, or markets, and to a much lesser extent, which incentives would be most effective at spurring more rapid and pervasive deployment of cleaner energy technologies.

In the United States, for example, it is common to hear members of Congress voicing concern about trade barriers and currency valuation in China. In global climate change negotiations, India routinely protests its perceived lack of access to advanced energy technologies. Developing countries criticize industrialized nations about their failure to transfer cleaner technologies and are emphatic about the paltry levels of financing to facilitate international technology transfer.

In the business community, we hear CEOs of multinational firms express concern about the threat of intellectual property infringement in developing countries, and especially in China. Corporate leaders also complain about nontariff barriers to trade, such as local content requirements and China's so-called indigenous innovation policies, which have been reported to give preferential treatment to Chinese firms and result in a lack of access to developing country markets. They worry about regulation and expropriation in the host countries, and also about their ability to access capital to support business expansion and related exports.

While there is less heated controversy per se about technology transfer in the academic literature, some of the conventional wisdom seems outmoded. Academic scholarship on international technology transfer typically describes a "north" to "south" trajectory, where an advanced industrialized country transfers an embodied, physical piece of equipment to a developing country—imagine a car battery, for example.

Conventional wisdom also assumes that government involvement is required, which usually results in a small-scale, development-aid, project-by-project approach, much like the model utilized by the Clean Development Mechanism (CDM), the main technology-transfer mechanism in the Kyoto Protocol to the UN Framework Convention on Climate Change (UNFCCC).[1]

Using China as a laboratory, this book empirically investigates these controversies and assumptions to determine the extent to which certain barriers and incentives are valid, at least in the Chinese context, and identifies the conditions necessary for motivating the international diffusion of cleaner energy technologies. The book also advances our understanding and updates the theory about the international transfer of cleaner and more efficient energy technologies. This introductory chapter provides a high-level overview of the main findings of the book, explains the rationale and need for the global diffusion of cleaner and more efficient energy technologies, defines "clean" and "technology," clarifies what is already known about the flow of cleaner energy technologies across borders, and supplies a detailed explanation of the research approach. At the end of the chapter, a guide to the entire book is provided.

Main Findings

Some may think of the barriers to technology transfer as a wall—perhaps most vividly the Great Wall of China. Size, however, is always a matter of perspective. It has been said that the Great Wall is the only human-made structure visible from space, but this belief was shaken when Yang Liwei, China's first astronaut, said he could not see it. The Great Wall is most easily detected through radar imagery. No human-made structure, much less the Great Wall, can be seen from the Moon (NASA 2005). At the global scale, then, even the Great Wall is not so big.

Moreover, we must remember that any wall is only as good as its sentries.[2] Who are the sentries? They are the protectors who strive to make the wall impregnable. Sentries who are alert and vigilant can make the wall effective in keeping technologies and ideas from getting through. As Manchurian and Mongolian tribespeople discovered, however, there are always ways to get around a wall. The search for entry can cause a long detour, cost money, and be troublesome, but it is almost always possible. Like water flowing downhill, profit-seeking individuals and firms find ways to penetrate or circumvent obstacles. They find cracks or openings, and guards who are either asleep or vulnerable to bribes or trickery.

This book begins, then, with the understanding that walls can be circumvented, and indeed clean energy technologies are penetrating all political borders on the planet, just not as fast or effectively as they could. In fact, it is not clear that a wall exists at all. Some of the oft-cited barriers to the diffusion of cleaner energy technologies present significant challenges, which are analyzed in great detail later in this book, but none are impenetrable.

This brings us to the first of four main insights in this book: the barriers are not nearly as daunting as many people assume; there is no "Great Wall" preventing the diffusion of clean and efficient energy technologies. These technologies regularly and frequently cross borders through a diversity of channels, including foreign direct investment, formation of joint ventures, licensing, consulting contracts, and joint research and development (R&D). Most of these methods are private, market-based mechanisms, although academia plays an important role as well by educating and training people from all over the world, and conducting research funded by domestic and foreign entities. Government creates the main incentive for cleaner technologies to move across borders, but private firms are the main agents of technology diffusion.

Second, the widespread concerns about intellectual property in the clean energy sector are not well founded based on empirical evidence from China. Industrialized countries like the United States and Japan are sensibly concerned about intellectual property infringement or outright theft in developing countries. Conversely, developing countries understandably worry about lack of access to intellectual property and how to get around patents filed defensively by existing firms to develop their own industries. The evidence presented in this book is that while there is indeed some infringement of cleaner energy technologies in China, it is much less widespread than typically assumed and largely a business challenge that can be effectively managed by careful firms. In fact, evidence emerges that many Chinese firms intend to play by the rules, and that the Chinese government is committed to the creation of a solid intellectual property rights protection regime. After all, as Chinese capabilities improve, they have intellectual property they want to protect too.

There is likewise limited evidence that developing countries have trouble accessing cleaner and more efficient energy technologies. To the contrary, some of the case studies in this book illustrate considerable ingenuity and resourcefulness on the part of firms in finding, acquiring, and further enhancing their technological capabilities through perfectly legal means. There are many different methods firms can use, and there is no one-size-fits-all approach. There are only two energy technologies for which evidence was found where there has been some withholding by advanced-country firms: hybrid-electric vehicle (HEV) technology and advanced natural gas turbines.

Third, the costs of advanced, cleaner energy technologies—and the availability of finance for paying the costs of such technologies—can certainly be a major barrier, but one that depends strongly on local conditions and policy. In China, there are virtually no barriers to accessing capital to support cleaner energy industries. In the United States, though, homeowners cannot easily get a loan for home energy retrofits, and US firms complain they cannot finance expansionary activities even when they can demonstrate market potential at home or abroad. In many countries, banks are unfamiliar with advanced energy technologies and therefore reluctant to lend money for them even when there are fast payback periods for more efficient technologies. Those banks that are willing to lend to clean energy firms may do so at higher interest rates, which increases the final cost of the technology to the consumer.

Of course, cleaner energy technologies cannot easily compete against incumbent, traditional energy technologies because the latter are less expensive, protected, or subsidized by governments, or benefit from long-term contracts. The fact that the full social costs of using fossil fuels are not incorporated into the prices of fossil fuel–derived energy puts cleaner energy technologies at a distinct disadvantage during the diffusion process (Brown and Sovacool 2011). If the costs of defending oil shipping lanes, contributions to trade deficits from importing foreign oil, the health effects of air pollution, and the impacts of climate change were included in the costs of using fossil fuels, their prices would be much higher and cleaner energy technologies would be able to more effectively compete.

The biggest barrier to the global commercialization of cleaner energy technologies is the failure of governments to create sensible policy incentive structures, at all levels of government—local, national, and global. Because the benefits of cleaner energy technologies mainly accrue to the public, private markets will never value them properly. It is therefore essential that government step in to compel the market to value environmental protection, human health, and energy security. For the global diffusion of cleaner energy technologies, therefore, governments must construct and mediate the market. This book will show how governments, at the local and national levels primarily, are already beginning to unleash global market forces.

Study of the Chinese experience with acquiring and selling cleaner and more efficient energy technologies provides two benefits. First, although China is unique in many ways, it has developed a model for how countries can acquire, modify, develop, manufacture, and export cleaner technologies. Second, examination of the Chinese experience also allows scholars to identify aspects of technology transfer theory that are no longer valid.

Green Mercantilism?

This book is mainly about the global diffusion of energy technologies, but it also contains the story of how China is transforming its energy system and developing a new industry. The growth of China's clean energy industry has been astonishing and unexpected, and has impacted the whole world. Some have welcomed China's entry into these markets, and others have viewed it with evident suspicion.

To some, China's prodigious effort to develop a cleaner energy industry is a case of "green mercantilism," where China disguises its motivation to increase its wealth and power through exports of cleaner technologies under the cloak of environmentalism.[3] Seeing substantial market opportunities around the world, China recognized it could dominate exports of cleaner technologies, accrue more foreign currency, and maintain its status as a banker to the world. Financial strength not only gives China economic power but also enables it to build up its military might. To the skeptics, the cleaner energy sector is just another industry that China aims to control, and any altruistic motives attributed to China are idealistic or apologist.

China's temporary export ban in 2010 on rare earth minerals to Japan is the best example of mercantilist behavior in the clean energy sector. The ban caused Japanese manufacturers to suffer because the minerals are used in many electronics and automotive applications. China accounts for 97 percent of rare earth exports, and half of China's exports go to Japan. Although China denied it had imposed an export restriction, the halt in exports coincided with an incident when Japan detained a Chinese fisherman who was near disputed islands in the East China Sea (Inoue 2010). Had the export restriction continued, damage to the Japanese economy would have been substantial and certain Japanese clean energy industries reliant on rare earths (e.g., HEV) might have been decimated.

Also, domestic subsidies that lower the cost of production and industrial policies that are aimed at fostering the industry can be interpreted as evidence of China's goal to dictate global trade in cleaner technologies. The US Commerce Department imposed tariffs on imports of Chinese solar PV modules in 2012 after making a determination that Chinese module producers were receiving illegal domestic subsidies that allowed Chinese firms to export at below-market prices, which in turn harmed US competitors. The Commerce Department could not determine the level of subsidization due to a lack of transparency so it assumed the worst and imposed high import tariffs that contributed to the bankruptcy of many Chinese firms.

While these arguments are persuasive, they do not explain other actions taken by the Chinese government. Why would the government announce a carbon tax on its own economy if it only intended to strengthen its economic might? A carbon tax would hit China harder than any other country because it is the largest producer and consumer of coal in the world, and coal is the most carbon-intensive fuel. Why would it establish expensive domestic feed-in tariff subsidies to accelerate the deployment of wind and solar technologies within China when it could continue producing electricity far less expensively with its relatively abundant coal reserves and export these technologies instead? The cost of electricity strongly affects China's industrial exporters because approximately two-thirds of Chinese electricity is consumed by the industrial sector. Why would China shut down hundreds of coal mines and prematurely retire hundreds of coal-fired power plants if its primary concern were energy security? Why would the twelfth five-year plan for the Chinese economy state that "China must set a reasonable limit on the total amount of national energy consumption" and "establish a carbon market step by step"? Why would China impose fuel-efficiency standards on its automobiles that are stricter than their US equivalents?

To understand these actions, one must recognize that the Chinese government indeed desires to create a sustainable economy in all senses of the word. Ever since the ninth five-year plan (1996–2000), the government has stated its intention to change the "development mode." Former premier Wen Jiabao is increasingly emphatic about the need for change at all levels of Chinese society. To provide one telling example, in an October 2011 teleconference with local officials he stated, "The party committees and governments at all levels must consider energy conservation and emission reduction as *the most important task* for promoting scientific development, as *the most important measure* for transforming the economic development pattern, and as *the most important index* for evaluating cadres at all levels" (Wen 2011; emphasis added).

Some regions of China are experiencing the worst levels of air pollution in the world. The multiple threats of climate change to China, especially reduced river runoff from the Tibetan plateau (and thus freshwater supply), sea level rise in the Pearl River delta, and increased flooding, all worry the Chinese government. In the development sense, transformation of China's economy from a resource-intensive, heavy manufacturing model to a less resource-intensive diversified economy is the government's main economic goal because more higher-quality jobs can be generated. To the Chinese government, therefore, the most important strategic effort is to reduce the resource intensity of the economy. Chinese foreign dependence on oil, iron ore, aluminum ore, copper, and other minerals already exceeds 50 percent. The fewer the resources consumed, the more efficient the economy, the more pollution is constrained, and the fewer resources that must be imported at great cost from abroad. China's oil import bill was $229 billion in 2011 (Tan 2012). Is the desire to be less dependent on foreign resources mercantilist? In the strict sense of the word it could be interpreted as such, but from a realpolitik perspective, it is understandable that the Chinese government would prefer to be less reliant on foreign oil and gas.

How does one explain China's investments into a clean energy industry if it is not motivated by mercantilism? China's energy consumption is already the largest in the world, having surpassed US energy consumption in 2009. China thus has the largest energy market in the world right inside its borders. While it could rely on foreign technology and imports to meet its domestic demand for cleaner energy, it would be forgoing a chance to profit in the market. Also, the Chinese government recognized that certain energy technologies were more expensive (especially solar PV) and that they would not succeed in the Chinese market without market-formation policies. China was later than most countries in creating feed-in tariffs for solar PV (the United States still has not done so), but after the costs of solar PV modules had fallen substantially, it finally did. If provision of low-interest loans and other forms of subsidization to Chinese PV manufacturers helped them to reduce the cost of solar PV, that is a good thing from the Chinese perspective because the costs were previously prohibitively high. The US government, at both the federal and state level, has also provided loan guarantees, investment and production tax credits, grants, and subsidized land. China's entry into this market has helped to bring solar PV within reach of consumers everywhere, and spurred the growth of solar installation firms in the United States and Europe as well as increased demand for the raw materials used in solar PV, largely sourced from outside China.

As in many cases, the truth is probably somewhere in between the two extremes. One cannot deny that the Chinese have an appetite for economic opportunity, they are fearsome manufacturers, and they have used an export growth development model for decades. On the other hand, there is plenty of evidence that the Chinese government is committed to transforming China's economy into one that is less resource intensive and environmentally sustainable. By establishing markets for cleaner energy technologies in China, the government has created demand for clean energy. It is understandable that Chinese firms want to meet this demand and that the government encourages them to do so at least cost. So long as foreign firms are allowed to fully participate in the Chinese market, this should be unobjectionable.

The Need for the Global Diffusion of Cleaner Energy Technologies

The more rapid and thorough diffusion of cleaner and more efficient energy technologies is urgently needed. Many rationales exist for spurring the global deployment of technologies such as solar PV, advanced batteries, LED lights, gas turbines, wind turbines, carbon capture and storage (CCS), geothermal, heat pumps, and hybrid cars. Motivations will differ among countries, individuals, and firms. In places where conventional energy resources are scarce, such as Japan, concern about oil and gas imports along with the resulting energy insecurity can be highly motivating. Japan's remarkable energy efficiency and its historically heavy reliance on nuclear power can be understood when one realizes that Japan has virtually no fossil fuel resources of its own. In other places, such as Germany, people value conservation and environmental protection highly, and they are willing to pay taxes on fossil fuels and subsidize renewable energy. Other countries or firms might be motivated by the desire to enhance their competitiveness and create jobs in an industry that is perceived to be more sustainable in all senses of the word. The United States and China are both examples of countries that want to compete in the so-called green economy. No country has only a single motivation but rather many, which interact with each other, and result in different national goals and strategies. A global diffusion of cleaner and more efficient energy technologies is highly desirable no matter the motivation, and indeed vital for human health, security, and economic prosperity.

Avoiding Climatic Disruption

Invisible "greenhouse" gases, which get their name from their ability to trap and retain heat in the atmosphere, alter the Earth's climate. There are dozens of greenhouse gases, including many chemicals such as chlorofluorocarbons (CFCs, the same culprit that caused the ozone hole) and sulfur hexafluoride (once used as the "air" in Nike Air tennis shoes), which has an "average" lifetime in the atmosphere of approximately 3,200 years.[4] Carbon dioxide (CO_2) is the most abundant greenhouse gas in the atmosphere today, and it is released into the air when fossil fuels are burned. Ever since the beginning of the Industrial Revolution when humankind began using coal, oil, and gas for its energy needs, a great quantity of excess CO_2 has accumulated in the atmosphere because the average residence time in the atmosphere of CO_2 is approximately 100 years.[5] In other words, humans are adding CO_2 more rapidly than natural processes such as absorption by plants, soils, and the oceans can remove it. One can imagine a bathtub into which water is pouring from the spout into the bathtub faster than the drain can take the water out and it soon overflows. The preindustrial concentration was about 280 parts per million (ppm), and as of the latest year available, the current concentration of CO_2 is 397 ppm and greenhouse gases as a whole have now surpassed 450 ppm of CO_2 equivalent.

Cleaner and more efficient energy technologies can reduce or virtually eliminate the emissions of greenhouse gases in several ways. First, energy-efficient technologies do more with less, reducing the demand for energy altogether. Meanwhile cleaner, low-carbon technologies can replace dirtier technologies, curtailing emissions per unit of energy used. The third option is to capture CO_2, and store it either in depleted oil and gas wells or deep saline aquifers—a process called CCS. Over time, inventors and entrepreneurs may discover ways to utilize the CO_2 for other applications. CCS is being demonstrated in a number of locations around the world at varying scales, but it is not yet a commercially available technology (IPCC 2005).

Countless books, scholarly journal articles, and studies have been written about the impacts of climate change, and let it suffice to say that the risks of climate change are great. Avoiding significant climatic disruption would save many low-lying coastal areas from sea level rise, help farmers adapt to different growing conditions and reductions in soil moisture, reduce the number of extreme weather events like floods, droughts, and heat waves, reduce the number of forest fires, and limit the spread of infectious diseases. The faster the rate of global emissions can be slowed, the greater the chance that climate disruption will be minimized, and that humans, animals, and plants will have time to adapt to the changes inevitable because of greenhouse gases already emitted into the atmosphere.

Economic Growth, Competitiveness, and Jobs

Some countries are already benefiting substantially from the development of alternative energy industries. In 2011, $263 billion was invested in clean energy around the world. The countries investing most intensively in a clean energy economy are, in order, the United States, China, Germany, Italy, the rest of the EU-27, India, the United Kingdom, Japan, Spain, and Brazil (Pew Charitable Trusts 2012).[6] These investments are helping these countries' industries to become significantly more competitive. China, Germany, and Japan are the top three exporters of solar PV and wind components in absolute terms, with China exporting US$10.8 billion, Germany US$6.9 billion, and Japan US$6.1 billion worth of clean energy technology on average annually between 2007 and 2009. Denmark and Germany are the two largest exporters of wind equipment, accounting for 22 and 19 percent, respectively, of global wind component equipment exports from 2007 to 2009. They are followed by China, India, Spain, and Japan. For shares of world solar PV component exports, China (29 percent), Japan (17 percent), and Germany (15 percent) hold the top three positions. There is quite a gap in the shares of exports held by these top three countries and the next three, which are the United States (6 percent), Korea (3 percent), and the United Kingdom (3 percent). Denmark is the only country where the export share of wind and solar energy technology accounts for more than 1 percent of the nation's total exports, however. In Denmark, exports of wind components alone account for 2.3 percent of the total national exports (Liebert 2011).[7]

Competition between the United States and China in the clean energy sector is sometimes characterized as a race. The political leadership of both countries is firmly committed to making clean energy a cornerstone of the economy. According to US president Barack Obama (2011), "As we recover from this recession, the transition to clean energy has the potential to grow our economy and create millions of jobs—but only if we accelerate that transition. Only if we seize the moment." Likewise, former Chinese premier Wen Jiabao (2009) remarked, "China ranked number one in the world in terms of installed hydro power capacity, nuclear power capacity under construction, the coverage of solar water heaters and cumulative installed photovoltaic power capacity, and fourth in the world for installed wind power capacity. These are major achievements in China's efforts to adjust economic structure and transform the development pattern. They also contributed positively to the global endeavor to develop green economy and tackle climate change." China has since become the global leader in installed wind capacity as well.

Significant economic benefit can be generated from clean energy technologies. In Germany, 367,000 jobs are estimated to derive from renewable energy as of 2010, compared with 943,200 in China, and 404,700 in the United States. In comparison, as of December 2011, there were 724,000 jobs in the US auto industry, not including dealers (Worldwatch Institute 2008; Federal Ministry for the Environment, Nature Conservation, and Nuclear Safety 2011; Bureau of Labor Statistics 2011).

Finally, the less a country has to spend importing fuel, the better its overall trade balance will be, which in turn can improve the overall health of its economy. The requirement that energy imports be paid in dollars is a particular burden for poorer developing countries that have difficulty earning hard currency to pay for imports. In 2011, the United States imported $332 billion in petroleum-related products, which was equivalent to 60 percent of the total US trade deficit that year (US Census 2012).

Improved Energy Security

The national security benefits of using cleaner and more efficient energy technologies can be substantial. The more a country can reduce its overall consumption of fossil fuels, the less that country has to import from other countries (if it does not already have these resources domestically). Although it is unlikely that any single country would be unable to import fuel even if it were boycotted by another country because the importing country could always turn to alternative suppliers, energy import dependence can make a country economically vulnerable to rising energy prices.

Reduced Air and Water Pollution

Anyone from a country like the United States or Sweden who travels to a rapidly industrializing country like China or India can be shocked by the visible air pollution in these countries. Non-CO_2 air pollution also comes from fossil fuel combustion. Actually, both China and India have made tremendous progress in addressing their air and water pollution problems, but there is still plenty of room for improvement. Of course, pollution used to be much more severe in the United States, Europe, and Japan, but due to increasingly stringent air and water environmental regulations enacted since the 1970s as well as robust enforcement regimes, the air and water are relatively clean in these regions.

These environmental regulations come in different forms, but the two main types are performance and environmental quality standards. With a performance standard, the factory, power plant, or automobile must emit no more than a certain amount of pollution for each unit of fuel consumed or service delivered (e.g., grams of carbon monoxide emitted per kilometer driven). Quality standards require that ambient air or water must contain less than a prescribed concentration of a particular pollutant. When these levels are exceeded, regulations may require fuel switching of power plants and factories to a cleaner fuel or technology, or in extreme cases to the halting of releases of the offending pollutant into the air or water until concentrations drop to an acceptable level.

In both cases, a company can decide which cleaner technologies to use in order to achieve a given level of performance. In fact, these regulations sometimes purposefully encourage companies to invent or improve pollution-control technologies to achieve cost reductions. Implementing environmental policies comes at some cost because the firms must acquire or develop the cleaner technologies, but this cost is reduced as such technologies become more readily available. These policies also provide public benefits including reduced pollution, which in turn can result in improved human health and less ecological damage. The benefits can also be more direct to the firms if they discover a cleaner process that actually results in cheaper production costs, and this in turn can improve their competitiveness in the industry overall. This idea is known as the "induced innovation" hypothesis, put forward originally by Michael Porter and Claas van der Linde in 1995. As they wrote, "Properly designed environmental standards can trigger innovation that may partially or more than fully offset the costs of complying with them" (Porter and van der Linde 1995, 98).

Definitions

There is no energy technology that is perfectly clean. Every single energy technology has at least one liability associated with it.[8] Coal, for example, is the most carbon-intensive of fuels, meaning the combustion of coal releases the most CO_2 of any fuel. Coal combustion also emits particulate matter, black carbon, toxic metals including mercury, and also sulfur dioxide and nitrogen oxides, which can cause acid rain and smog. Yet emissions from coal can be greatly reduced. Some technologies like CCS can theoretically remove upward of 90 percent of the CO_2 emissions from coal. And other technologies like integrated gasification combined cycle (IGCC) remove virtually all the conventional air pollutants, including mercury. Oil production can result in major oil spills in the ocean and on land, and oil combustion causes urban air pollution and climate change. For the fossil fuels, it used to be generally thought that natural gas is the most environmentally benign because of lower CO_2 and sulfur dioxide emissions, but again, much depends on how each fuel is mined or extracted, how much leakage there is of the fuel during extraction, processing, and refinement, and which technologies are employed in production and consumption to reduce the environmental impacts. Now that natural gas is being extracted from shale formations, environmental concerns about natural gas are rising due to the potential impact of shale gas extraction on local water supplies and also the significant potential leakage of methane into the atmosphere because methane is a potent greenhouse gas. Hydroelectric electricity does not emit air pollution, but it is generated from turbines installed in dams, which often cause major local ecological and human disruption. Wind turbines can be noisy and disruptive of bird wildlife if placed amid migratory routes. Depending on how they are produced, solar PV modules can cause occupational hazards for factory workers due to exposure to toxic substances. Nuclear energy technologies emit hardly any air pollution, but they do consume large quantities of water (especially for fuel enrichment and cooling), can create large volumes of radioactive waste depending on the precise technology employed, and depending on the particular technology, can produce weapons-grade material that can be stolen by terrorists to produce bombs (Allison 2004; Bunn 2010). Biofuels production can cause hazardous waste runoff as well as emissions of nitrous oxide and significant net amounts of CO_2. Their impacts also depend greatly on what existed on the land before it was cultivated, the type of crop being produced, the local conditions, the amount of fertilizer used on the crops, the energy consumed on the farm, how far the crops must be transported to refineries, and the fuel use of the refineries.

Therefore, no technology can be considered a silver bullet solution even if it is the cheapest, cleanest, most efficient, or most "secure" energy source available. The World Energy Assessment of 2000 provided a ranking of human-generated environmental insults by sector based on a human disruption index (the ratio of human-generated flow to the natural baseline flow), which showed that all sectors contribute to the human disruption of the natural environment. As of the mid-1990s, the top five insults were lead emissions to the atmosphere from fossil fuel burning and manufacturing; oil added to the oceans due to petroleum extraction, processing, and transport as well as disposal of oil wastes; cadmium emissions to the atmosphere due to fossil fuel burning, agricultural burning, and metals processing; sulfur emissions to the atmosphere due mainly to fossil fuel burning; and methane flow to the atmosphere mainly due to agriculture (rice paddies and domestic animals), fossil fuel burning, and landfill emissions (Holdren and Smith 2000).

Despite the fact that no energy technology is truly clean, some technologies are much cleaner and more efficient than others. Technologies that improve the efficiency of energy use are at the top of the list because they reduce energy consumption overall, which in turn causes fewer pollutants to be released into the air and requires less energy to be imported from other countries.[9] Solar, wind, geothermal, and other renewable energy technologies are also relatively clean. It is important to consider the health and environmental impacts of the full "life cycle" of all energy technologies, including mining or extraction, processing, production, manufacturing, refining, distribution, and the use of the different energy technologies. Determining the life cycle impacts of each technology is difficult and highly context dependent.

Frequently, the phrase "cleaner energy technologies" will be used in the book, and it is meant to include more efficient, renewable, and low-carbon energy technologies. Again, the relative cleanliness of a given energy technology is based largely on the local context, and this context is set primarily by governments that (may) regulate the extraction, processing, production, and use of various energy technologies. A central message of this book is indeed that government policy is a prerequisite for the global diffusion of cleaner energy technologies. It is generally assumed that efficiency technologies are quite clean, renewable energy technologies fairly clean, and advanced fossil-fuel-based technologies like combined cycle gas, combined heat and power, and CCS are relatively clean compared with conventional coal-fired power plants. As already noted, the focus of the book is not to promote or advocate individual technologies but rather to determine how to better facilitate the global diffusion of cleaner energy technologies.

As for the question of what is meant by an energy "technology," the definition advanced by W. Brian Arthur is helpful: a technology is "a means to fulfill a human purpose." It can be a method, process, or device, and also an assemblage of practices and components. Many consider technology to be knowledge—know-what, know-how, and know-why (Garud 1997). Although Arthur (2009, 218n27) disagrees that "technology is knowledge," he concedes that knowledge is necessary for technological change, which is the focus of this book.

The third frequently used term in this book is "globalization." It is reluctantly used because the term lacks a precise definition, but for our purposes, one must be provided. Globalization is the economic integration of an industry across borders through trade, government policy, capital flows, international institutions, and labor migration. As Ed Steinfeld (2010, 22) writes, "Globalization is not really about different parts of the world trading at arm's length and competing head to head. Quite to the contrary, it is about the world, including places once considered the farthest frontiers, getting pulled into complex production hierarchies that once existed only within the firm." This book explains how the clean energy industry became globalized during the 1990s and 2000s.

Innovation and the Flow of Cleaner Energy Technologies across Borders

The diffusion of technology can be thought of as part of the final "stage" in the growth or life cycle model of innovation. This model was originally conceived as beginning with invention through R&D. Technologies were born through invention, and then they entered adolescence through a demonstration stage, where they were experimented with through much trial and error, at different scales. If they made it to adulthood, technologies would then be deployed through a process of diffusion. Once a technology had been widely diffused and the market was saturated, its life was more or less over. This stylized model of innovation was originally thought to be orderly and linear, starting with invention and concluding with diffusion. Over time, the dynamism and messiness of the innovation process began to be observed. Multiple feedbacks and interactions were found to exist. One example is that when technologies are demonstrated, they often need to go back to the R&D phase for further refinement. During the so-called deployment phase, technologies are placed into different contexts, and frequently must be adapted for that context, which in turn can require demonstration or adaptation of the technology for that market. Through this adaptation process, new technologies can be discovered, because after all, technological innovation is a cumulative process that includes both incremental and radical changes to existing technology. Most important, innovation does not occur in isolation from society, but rather it is strongly affected by the social system, including different actors, networks, and institutions (Rogers 1995). Various corresponding incentive structures affect innovation, including the market and government policies (Gallagher et al. 2012). Knowledge and learning are key aspects of any innovation system, and uncertainty is always prevalent.

In short, it is now clear that innovation is not just "invention" or "R&D" but rather a set of processes that as a whole should be thought of as an innovation *system* (Carlsson and Stankiewicz 1995). Technological innovation systems are understood to encompass all the factors that influence an innovation outcome. There are many different functions in an innovation system, including entrepreneurial activities, knowledge development, knowledge diffusion through networks, guidance of the search, market formation, resource mobilization, and creation of legitimacy/counteraction to resistance to change. The process of entrepreneurship brings new technologies, knowledge, and networks to the market. Knowledge development refers to learning and experimentation. Knowledge diffusion through networks includes the exchange of information among actors and spillovers from one industry to another. "Guidance of the search" refers to choices that actors make about how to focus innovation efforts. Although "picking winners" is much disparaged as a technology policy tool, the fact is that resources are always limited and so priorities must be established to guide the innovation process. Market formation is the creation of protected space for new technologies to create niche markets or correct for market externalities. Human and financial resources must be mobilized within an innovation system to produce desired outcomes. Creation of legitimacy refers to the fact that incumbent firms and technologies rarely welcome disruptive technology, often resist it, and in many cases may attempt to block its introduction. These functions help explain why the innovation process is so dynamic—they all interact and can be supported (or weakened) by government policy (Hekkert and Negro 2009). Overcoming the status quo and the policies that congeal around it to protect it has to be a major function of governmental policy if innovation is to thrive.

The systems approach to innovation permits the conceptualization of different types of innovation systems—national, regional, and sectoral innovation systems (for a review, see Lundvall 2007). Each conceptual approach emphasizes different system boundaries, whether geopolitical (e.g., nation-state borders or regional groups) or economic (e.g., industry segment). This book focuses mainly on the interactions of two types of innovation systems: national innovation systems (NISs) and the global energy technology innovation system (ETIS). As such, both concepts must be defined.

Many scholars developed the NIS concept (e.g., Giovanni Dosi, Charles Edquist, Christopher Freeman, Bengt-Åke Lundvall, Richard R. Nelson, Gerald Silverberg, and Luc Soete, among others). The NIS approach underscores all the aspects of an innovation system identified above, but concentrates on the nation-state as the main level of analysis, which stresses the role of national government and national institutions, such as "national" academies of science, education systems, ministries of science and technology, and so forth. As countries have developed economically, national industrial policy, the fostering of national champion firms, and the improvement of national education systems have all proven to be important. National innovation systems are of course open to international influences, but the national level of analysis allows for interesting and useful comparative study between countries.

The idea of an ETIS was developed more recently (Grübler et al. 2012), and it applies the systems approach to energy innovation. The ETIS is "the application of a systemic perspective on innovation to energy technologies comprising all aspects of energy systems (supply and demand); all stages of the technology development cycle; as well as all innovation processes, feedbacks, actors, institutions, and networks" (Gallagher et al. 2012, 139). It is a sectoral innovation system, with emphasis on technical change within the energy system. The ETIS can be applied to both national and global levels of analysis. In this book, the energy innovation system of China is examined in great detail as part of an NIS as well as a global innovation system for energy. Indeed, China's contribution to the global ETIS is already large and growing rapidly.

We can therefore think of technology transfer as part of the diffusion process within a broader innovation system. Technology transfer in the clean and efficient energy sectors necessarily bridges at least two NISs (and often more) within a global ETIS. Table 1.1 provides a typology of mechanisms for the cross-border flow of cleaner energy technologies. Here, it can be observed that most of the modes for international technology transfer are in the domain of the private sector. Some of the modes of cross-border transfer identified in table 1.1 are more conventionally

Table 1.1
Typology of mechanisms for the global diffusion of cleaner energy technologies

Mechanism	Variation(s)
Exports or imports of final goods	Equipment for manufacturing
Licenses	
Purchase of a foreign firm to acquire technology	Merger or acquisition
International strategic alliances or joint ventures	Partial or wholly owned
Migration of people for work or education	Entrepreneur, financier, consultant, or a employee who has worked or been educated in another country
Contract with a foreign research entity	Could be a contract with a university lab, government lab, or for-profit firm
Collaborative R&D	Research partnerships with foreign entities with shared intellectual property (IP) arrangements
Open sources	Exhibitions, conferences, books, papers, and patent documents
Bi- or multilateral technology agreements among governments for R&D	Could include private participation; may include support for capacity building or "tied aid"

Sources: Mowrey and Oxley 1997; Lanjouw and Mody 1996; Gallagher 2006a; Barton 2007; Lewis 2007; Odagiri et al. 2010; Lema and Lema 2010.

associated with environmental technology transfer (e.g., exports of embodied technologies), whereas others are less conventional in the clean energy industry (e.g., acquisition of foreign firms).[10]

Theoretical Contributions of This Book

This book provides considerable evidence that countries, developing or industrialized, can acquire, modify, develop, manufacture, and export cleaner energy technologies. The first conclusion is that the clean energy sector experienced a pronounced globalization of both the development and deployment of cleaner energy technologies during the early years of the twenty-first century. The main drivers of this globalization appear to be the internationalization of postsecondary education, ease and increased normalcy of migration, establishment of new national and subnational policies to boost energy efficiency and reduce pollution, creation of international institutions that facilitate cross-border investments and trade, and availability of capital in multiple countries to finance the cross-border movement of technology.

A number of aspects of technology transfer theory no longer fit well with the emerging evidence presented in this book about the cross-border movement of clean energy technologies. First, international technology transfer in clean energy is not a unidirectional north-to-south process anymore (Brewer 2008). It is north to south, south to north, south to south, and north to north. In fact, the NISs for clean energy no longer seem as relevant as the *global* ETIS that is observed in this sector. Global learning networks already exist, and firms are adopting a decisively global perspective on where they source and sell their technology, products, and services. Clean energy technologies are now rarely, if ever, developed in a single country and then physically transferred to another. Technologies are acquired from all corners of the globe using a diverse set of mechanisms, including licensing, hiring of talented individuals, purchasing foreign firms, and joint research contracts. Technology is cumulative and iterative, and now takes advantage of a global knowledge base. Entrepreneurs, multinational and domestic firms, and top research institutions including universities are highly global in their orientation. It is not that they do not respect international borders but rather that they have a global perspective.

Second, it is not at all clear that smaller, individualized energy technologies are better for developing countries. While the "small is beautiful" concept (Schumacher 1973) is attractive because it stands to reason that technologies must be appropriate for individual needs and conditions, it connotes an approach to technology transfer that is first, north to south in concept as well as almost *artisanal*—the interpretation is that technologies require extensive adaption and modification, and then must be transferred one at a time. There is little evidence that such a small and piecemeal approach will deliver the scale of diffusion that is required. In the context of climate change, it could be argued that small is imprudent and bigger is better. Cleaner energy technologies cannot be globally diffused at a scale with government-sponsored individualized technology transfer initiatives, project by project. Rather, private markets must be developed and harnessed to deliver the cleaner energy technologies where they are needed. Many "base-of-the-pyramid" markets are already benefiting from the rapidly declining costs of solar PV technologies, for example, that resulted from this globalized approach.

Bigger market scale is critically important to the global diffusion of cleaner energy technologies. Given pervasive market failures, the demand for cleaner energy technologies is mainly created through market-formation policies at the national and subnational levels. With greater demand and increasing standardization, more commercial experimentation will occur in diverse markets around the world. Industrial clusters for mass production will emerge, competition will increase, and costs are likely to fall due to economies of scale, learning, or other reasons. These market-formation policies are required because the "natural" market rarely values all the benefits of cleaner energy technologies (Ockwell and Mallett 2012). These benefits can include reduced harm to the land and ecosystem, reduced air and water pollution (which leads to improved human health), improved energy security, and so forth. Other market failures and distortions are discussed in chapter 6. Market-formation policy thus necessarily goes well beyond the creation of niche markets, which are by definition too narrow to achieve the scale of the market that is required. Yet market-formation policy need not last indefinitely. If sufficient demand for a cleaner energy technology is generated for whatever reason, then the policy can and should be modified or eliminated. For instance, many feed-in tariff policies are designed to subsidize the cost of cleaner electricity generation technologies like solar PV. As the cost of solar PV wafers has come down, the gap between the cost of generating electricity from solar PV versus conventional coal or gas has narrowed. When the feed-in tariffs were not reduced accordingly, solar PV producers were oversubsidized. A tension thus exists between the durability and flexibility of market-formation policies (Carlson and Fri 2013). A more complete discussion of the theoretical implications is provided in chapter 7, where an integrated theory of the diffusion of clean energy technologies is also offered.

The Research Approach

The main research questions for this book are: Which barriers most inhibit the global diffusion of cleaner and more efficient energy technologies, and, Which incentives or conditions are necessary to motivate the global diffusion of these technologies? To answer these questions comprehensively, a mixed methods approach was employed depending on the component of the overarching research question that was under examination in each chapter. Several quantitative databases were constructed and analyzed, nearly one hundred in-depth interviews were conducted with key actors during field research in China, Germany, and the United States, and the findings were juxtaposed with the vast literature on technology diffusion in general and cleaner technology development in China in particular.

This research largely employs a case study approach to identifying the barriers and incentives to cleaner technology diffusion in the world economy. While there is a growing tendency for large-N quantitative studies in the literature, the results of that literature remain ambiguous. Even though such approaches can yield considerable insight, they tend to tilt thinking toward causal factors that are readily measurable and neglect those factors—and their feedbacks—that are more difficult to quantify. While I do construct and analyze quantitative data for this work, where appropriate, I largely utilized case studies in order to trace the process of technology diffusion and disentangle the complexities of the causal mechanisms at play. The main advantage of this case study approach is the ability to examine and reconstruct the processes of cleaner technology acquisition and diffusion in the societal as well as institutional contexts in which technology was being developed and deployed. By conducting detailed case studies, it was possible to identify causal factors that prevailed (or not) across different subsectors of the clean energy industry.

Four new case studies presented in chapter 3 and drawn on throughout the book are the main empirical basis for analysis. These studies examine how four clean energy industries in China managed international technology transfer, including how foreign firms transferred technology into China, how Chinese firms acquired and developed technology, and how Chinese firms exported technologies. The four cases are gas turbines, solar PV, coal gasification, and advanced batteries for motor vehicles. The choice of case studies is discussed in detail in chapter 3, but in short, at the outset the cases appeared to represent both "successful" (solar PV) and "failed" (gas turbines) technology transfer, and some in between (advanced batteries for vehicles and coal gasification). The batteries case allowed for analysis of an end-use technology in contrast to the other three supply-side technologies. The coal gasification and batteries cases seemed to have relied more on indigenous development versus technological acquisition from abroad (solar PV and gas turbines). In the end, not all of these assumptions were correct, as the reader will discover. The development of China's wind industry is an extremely interesting case study, too, but Joanna Lewis (2007) and Ru et al. (2012)

have already laid a solid foundation, and their findings are contrasted with the new case studies. In future research, additional case studies of end-use technologies would be especially useful as there is a major gap in the literature in this respect.

The case studies were constructed through dozens of in-depth semi-structured interviews with individuals in firms, academic experts, and government officials as well as from articles and books in the scholarly literature and popular press. To more easily visit factories, interview people within the firms in the sector, and interview experts and government officials in China, I was based at Tsinghua University in Beijing as a visiting professor at the School of Public Policy and Management in 2010. I also made separate research trips to China, and conducted interviews in China through Skype and by telephone. To get the non-Chinese side of the story, I primarily relied on interviews in the United States and Germany that were completed mostly during 2011 and 2012. I made two trips to Germany, one during summer 2011, and a second in 2012 when I was in residence at the American Academy of Berlin as an EADS Distinguished Visitor. All in all, I conducted nearly one hundred interviews. A list of those interviewed and consulted is available in appendix A. In addition to the four new case studies, I compiled a database of existing case studies of clean energy technology transfer mainly from the scholarly literature, but also including "gray," non-peer-reviewed, published studies. Twenty-seven scholarly case studies of the international transfer of cleaner energy technologies and seventeen other case studies were identified to enable comparative analysis. A list of the case studies is available in appendix B. Most Chinese sources did not wish to be directly quoted with attribution so many interviews are cited by number and date, not name.

I also structured focus groups to gain insight on these matters. In January 2012, I organized a workshop on the globalization of clean energy technologies at the Fletcher School at Tufts University, where about twenty-five private sector representatives and academic experts were invited to share their direct experience with the international transfer of cleaner energy technology, with emphasis on transactions that did not include China. One of the goals of the workshop was to better understand what was unique to China and what could be considered more generalized experience with the transfer of cleaner energy technologies. A second goal was to understand what was unique to cleaner energy technologies as compared with other industries such as pharmaceuticals or chemicals.[11]

For the chapter on intellectual property, in addition to the case studies, I conducted an analysis with Amos Irwin of clean energy patents granted by China's State Intellectual Property Organization (SIPO) to determine the rate at which both Chinese and foreign firms were filing different kinds of patents. A database of all Chinese and foreign patents granted in China for ten different clean energy technologies between 1995 and 2011 was developed, and used as the basis for analysis. A Chinese government database of IP-related court cases was also used to determine the extent to which patent infringement was occurring in the four cases studied, how frequently foreign firms were involved, and if so, what the outcomes were. The results from both of these efforts are presented in chapter 5.

Finally, extensive comparative policy analysis was conducted in order to understand the policies being used by different countries to spur the development and/or enhance the competitiveness of their clean energy industries, and also accelerate the deployment of these technologies. I analyzed market-formation policies, industrial or manufacturing policies designed to strengthen the clean energy sector, technology and innovation policy, and export-promotion policies. A timeline of major market-formation policies was constructed (see appendix D) and compared with changes in the volume of global trade in clean energy technologies to assess the strength of the relationship between the two factors. Many studies exist in the scholarly literature about the efficacy of these policies, which were helpful. To fill in the gaps about certain policies and their effects in the literature, I relied on interviews with government officials and firms. This analysis is presented in chapter 4.

Guide to the Book

This book is analytically organized around the three main hypothesized barriers to the international transfer of technology with chapters on policy ("The Essential Role of Policy"), intellectual property ("No Risk, No Reward"), and cost and finance ("Competing against Incumbents"). Before these chapters, however, we go "Into the Dragon's Den" (chapter 2) to understand why China is a unique and essential laboratory for examination of the global diffusion of cleaner energy technologies. China's energy and economic development is explained, and the government's many policies related to energy and environment are reviewed. China's innovation system for energy is explained in this chapter as well. In chapter 3, "Four Telling Tales" are related through the case studies on gas turbines, advanced batteries for cleaner vehicles, solar PV, and coal gasification for electricity production. Chapter 3 presents the case studies as complete stories, and based on them, then provides a high-level analysis of the key barriers and incentives to the global diffusion of cleaner technologies. Chapters 4 through 6 analyze in detail three of the barriers that seemed most daunting as discussed above—policy, intellectual property, and cost and finance. Detail from the case studies is drawn out in these topical chapters. Chapter 7 concludes the book, first with an insight about the apparent globalization of the cleaner energy industry beginning around the turn of the century and what is driving this globalization process, and then compares the evidence presented in the case studies with existing theories about international technology transfer. Finally, a new integrated theory of the global diffusion of cleaner energy technologies is offered, and the policy implications of the research findings are provided.

This book was originally intended to both update and enhance our understanding about technology transfer for cleaner energy technologies and empirically examine the extent to which certain barriers really exist for the global diffusion of cleaner energy technologies. The research led to the conclusion that certain forces of globalization are already contributing to the global diffusion of cleaner energy technologies. The most important finding is that national and subnational policies are key to unleashing the market forces that have the potential to greatly improve the human and natural condition through the deployment of cleaner energy technologies around the world.

2

Into the Dragon's Den

In chapter 1, the rationales for a more pervasive and rapid global diffusion of cleaner and more efficient technologies were explored, with emphasis on the need to reduce the threat of global climate change, reduce conventional air and water pollution, enhance energy security, and improve economic prosperity. In aggregate, but not per capita terms, the top three greenhouse gas emitters are China, the United States, and the European Union (in that order). As the largest emitter, China is arguably the most important country in which to examine the cross-border diffusion of cleaner and more efficient energy technologies.

China as a Laboratory

China is intrinsically interesting as an enormous country with 20 percent of the world's population, the second-largest economy, the largest energy production and consumption, and the largest greenhouse gas emissions on the planet. China is also worth examining because of the polarizing perceptions many have about this rapidly industrializing country. To a minority, it represents a monstrous threat that must be contained or even fought. To others, it is the new "Western" frontier—a land of opportunity and riches. Because China is now the largest clean energy market in the world, and despite (or perhaps precisely because of) the emotional reactions many have about China, this book goes into the dragon's den to empirically explore cross-border transfers of cleaner energy technologies.

There are problems with focusing on China as the locus of inquiry, however. China is a large, almost infinitely complex place, and as such, it is usually impossible to generalize about China. In addition, China has many unusual characteristics not shared with any other country in the world. First, China's energy system is strongly affected by its particular natural endowments, especially its historically vast, though quickly diminishing, coal reserves. Only two countries have larger coal reserves: the United States and Russia. Still, China's reserve-to-production ratio is only thirty-five years, much lower than either of the other two countries, which is one of the reasons it became a net importer of coal in 2009 (BP 2011). Second, the political economy of China is unlike that of any other country. It is a thriving "socialist market" economy where capitalist forces of competition are fierce and brutal, but also where state ownership of firms is common, and the government frequently intervenes in markets. It is also a place where more than 650 million have been lifted out of extreme poverty in the past thirty years, but income inequality has become startlingly high. The Gini coefficient (a measurement of income inequality) for China increased from 41 in 2007 to 48 in 2009, which places China in between the Dominican Republic and Madagascar (CIA 2012). The per capita income is $7,600, yet 200 million Chinese people still live on less than US$1.25 dollar a day (World Bank 2012).

China is also unique in the incredible availability of capital there. Ease of financing is a key reason why the Chinese have been able to acquire cleaner and more efficient energy technologies. As Justin Yifu Lin (2012, 207), former chief economist of the World Bank, writes in his book on the Chinese economy, "Technological innovation is not a free lunch, it needs capital." Understanding the sources of finance for clean energy technologies in China was not as much a focus of this book as whether or not the availability of finance posed a barrier or incentive to the cross-border movement of technologies. During the course of the research it became clear that the availability of finance was a major incentive. More research is needed to understand exactly how the Chinese government helped to finance this industry. The Chinese government has amassed huge financial resources that it can devote to fostering new industries, and it has committed to doing just that for what it calls "new" energy technologies. China's top four banks account for the majority of official lending in China, but informal markets and funds managed by local governments (for instance, local investment corporations or investment funds) are large. One estimate of the size of these nonbank sources of investment is that they equal about 4 percent of GDP annually, about half as much as formal bank lending (Naughton 2007). The total loans from the four largest state-owned national banks in China in 2010 equaled 203,793 100 million *renminbi* (RMB; US$3 trillion at 2012 exchange rates) according to statistics from the People's Bank of China.

Table 2.1
Total investment as a percent of GDP (2011)

China	48%
India	34%
Brazil	21%
Japan	20%
Germany	18%
United States	16%

Source: International Monetary Fund, World Economic Outlook Database, Washington, DC, http://www.imf.org/external/pubs/ft/weo/disclaim.htm, accessed April 2012.

Household deposits accounted for 43 percent of the sources used by financial institutions (People's Bank of China 2010). The total investment in China as a percentage of GDP is much higher than in the other major economies (see table 2.1). China's policy banks, especially its national development bank and Ex-Im bank, provide major support as well. In addition, China is increasingly attracting venture capital from around the world. Research for the case studies in this book indicates that local governments are crucial sources of capital for clean energy enterprises in China.

The Chinese energy market is already the biggest in the world, which presents gigantic challenges and opportunities for the Chinese government. The main challenge, of course, is how to satisfy the energy demand of the Chinese economy in a way that does not continue to damage the local environment, harm human health, or irreversibly alter the global climate. The opportunity is to use the lure of the market to motivate firms to contribute to the achievement of the social goals set by the government. No other country in the world has this unique set of characteristics, and so subsequent chapters in this book will weave evidence from China together with the larger body of scholarly literature from around the world.

This chapter begins by examining how economic development and energy are intertwined in China. Then the Chinese energy and environmental landscape is sketched, with an emphasis on how it has changed since Deng Xiaoping launched China's experiment with gradual economic reform and openness. Chinese government policies related to energy, climate change, and innovation are analyzed, and the drivers of these policies are explained.

Energy and Economic Development in China

Since 1978, China's economy has grown rapidly, with GDP increasing from 0.36 to 39.8 trillion RMB in 2010.[1] The average annual growth rate during this period was about 10 percent (National Bureau of Statistics 2011). In 2010, China's GDP surpassed Japan's and became the second largest in the world (World Bank 2012). Though China's traditional development model resulted in many achievements, it also brought about many serious problems, and is often characterized as imbalanced, uncoordinated, and unsustainable. Growth is depicted as being marked by high investment, high consumption, and high emissions. China's economy is one of the most resource intensive in the world, and its dependence on foreign oil, iron ore, aluminum ore, copper, and many other minerals already exceeds 50 percent (Xuan and Gallagher 2013).

Social development has taken second place to economic growth. The ratio of labor income to GDP has decreased consistently from 51.4 percent in 1995 to 39.7 percent in 2007. This means that the benefits of economic growth are accruing to enterprises and the government, but less so to labor. The traditional economic model puts the major stress on growth, which has led to insufficient fiscal expenditure on public services for citizens and lack of a social safety net (Wong 2009). A related problem is widening inequality, between rich and poor as well as urban and rural.

The imbalance among investment, consumption, and exports has created thorny tensions. The Chinese growth model has relied heavily on investment and exports, not domestic consumption. Currently, China has the highest investment ratio and the lowest consumption ratio in the world. Energy prices were subsidized to reduce costs for industry, which created a perverse incentive to use more energy, and not to conserve. Prices are higher now, but much lower than in other countries that tax energy or CO_2 emissions. To summarize, in China's traditional development model, China had abundant labor, plenty of low-cost capital, below-market prices of raw materials, a relaxed regulatory environment, a strong industrial base, and continuously improved infrastructure.

For years, the central government has been trying to reform the growth model. Michael Rock and David Angel (2005) use the term "policy integration" for China's process of linking environmental protection policies with policies for technological upgrading and the open economy. Indeed, the government has recognized that the resource intensity of the Chinese economy is not sustainable economically or environmentally. Export markets have weakened with the global financial crisis. Local competitive pressures are driving Chinese economic growth faster than they should from the standpoint of the national interest. As early as the ninth five-year plan (1996–2000), China's central government proposed shifting economic growth from an "extensive" to an "intensive" mode, meaning that growth should be driven by efficiency gains and technological innovation rather than material input. In 2003, the "scientific concept of development" was put forward by the central government. This scientific concept was predicated on more balanced development between urban and rural areas, among different regions, with more emphasis on environmental protection and social development.

Unfortunately, meeting these lofty goals has been much harder than anticipated. China's unchanged strong reliance on heavy industry has driven up energy consumption. Export markets are disappearing, and along with them go China's trade surpluses. In 2011, China's current-account surplus was only 2.8 percent compared with 10 percent in 2010 (World Bank 2012; Porter 2012, B1). Subnational levels of government have resisted reforms because they believe that economic growth is the number one goal, and each province and municipality wants to stand out with faster growth, so these subnational governments often support new industrial developments that are redundant with each other.

Energy consumption is directly tied to economic development, and the relationship between energy use and economic growth matters greatly in China. Although China increased its GDP fortyfold between 1980 and 2010, it did so while only quintupling the amount of energy it consumed from 1980 to 2009. This allowed China's energy intensity (ratio of energy consumption to GDP) and consequently the emissions intensity (ratio of CO_2-equivalent emissions to GDP) of its economy to decline sharply, marking a dramatic achievement in energy intensity gains not paralleled in any other country at a similar stage of industrialization. As can be seen in figure 2.1, the energy intensity of the Chinese economy was 3.4 tonnes of coal equivalent (tce)/10,000 RMB (2005 constant RMB) in 1980, and 1.08 tce/10,000 RMB in 2009. This achievement has important implications not just for China's economic growth trajectory but also for the total quantity of China's energy-related pollution. Reducing the total quantity of energy consumed also contributes to the country's energy security. Without this reduction in the energy intensity of the economy, China would have used more than three times the energy that it did during this period (Xuan and Gallagher 2013).

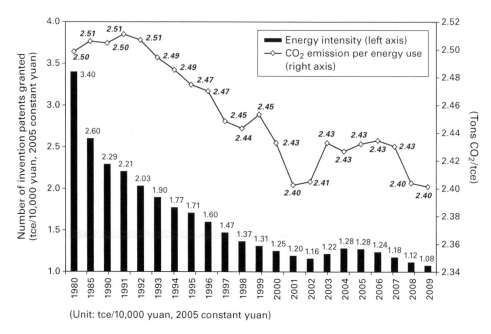

(Unit: tce/10,000 yuan, 2005 constant yuan)

Figure 2.1
Chinese energy and CO_2 intensities
Source: Xuan and Gallagher 2013; data from National Bureau of Statistics in China

The beginning of the twenty-first century brought new challenges to the relationship among energy consumption, emissions, and economic growth in China. Starting in 2002, China's declining energy intensity trend reversed, and the growth of energy use surpassed economic growth for the first time in decades. This trend continued until 2005. This reversal had dramatic implications for energy security and greenhouse gas emissions in China. In 2007, China's emissions were up 8 percent from the previous year, making China the largest national emitter in the world for the first time, surpassing US emissions that year by 14 percent.

The Chinese economy has grown every year since 1978 and undergone tremendous change. Many state-owned enterprises have been privatized, and many others have been reformed. Lin (2012, 74) has argued that under its planned economy, China initially pursued a "comparative-advantage defying" strategy by focusing strongly on large-scale, capital-intensive, heavy industries, but once China moved to a "comparative-advantage following" strategy, the economy flourished. During the reform period, industrial structure has changed a lot. Agriculture's share of GDP fell from 38 percent in 1980 to 9.4 percent in 2009, industry's rose from 27 to 42 percent, and services from 30 to 42 percent. Every sector of China's economy experienced substantial improvements in energy intensity, but the energy intensity of the industrial sector is still eight times that of agriculture and construction, and three times as large as services. Still, improvements in industrial energy efficiency have contributed to approximately three-quarters of the overall improvement in energy intensity of the Chinese economy (Xuan and Gallagher 2013).

Industry accounts for about 70 percent of China's annual energy consumption, and in turn, China's industrial base supplies much of the world (National Bureau of Statistics 2011). As a result, China's energy and environmental challenges are fueled in part by the global demand for its products. According to one estimate, one-third of China's energy consumption was export driven as of 2007 (Z. Li 2012). China today produces about 35 percent of the world's steel and 28 percent of its aluminum, for example, up from 12 and 8 percent, respectively, a decade ago. The centerpiece of the eleventh five-year plan was to promote the service industries—the so-called tertiary sector—because of their higher value-added to the economy, and the energy and environmental benefits associated with a weaker reliance on heavy manufacturing. The tenth five-year plan (2001–2005) strongly highlighted the need to shift to lighter industries, setting targets for the "secondary" and "tertiary" industries (which include lighter industry and services), and similar language has been included in subsequent five-year plans since then.

By shifting toward lighter industry, the composition of China's industrial structure would change, and in so doing, energy use should decline, and environmental quality should improve. The recent resurgence in heavy industry in China, responsible for the rapid emissions growth in the past few years, illustrates the challenge of facilitating this transition. In the twelfth five-year plan (see table 2.2), the government more explicitly identifies a new set of high-value strategic industries such as biotechnology and information technology, and also identifies energy saving, environmental protection, and new energy technology industries as essential to the future of the Chinese economy.

China's long-term energy security is dependent not only on having sufficient supplies of energy to sustain its incredible rate of economic growth but also on being able to manage the growth in energy demand without causing intolerable environmental damage.

Table 2.2
Key targets in China's twelfth five-year plan (2011–2015)

Percentage of nonfossil fuel in primary energy consumption	11.4%
Carbon intensity (CO_2/GDP) reduction	17%
Energy intensity (energy/GDP) reduction	16%
Strategic emerging industries	Energy saving and environmental protection, new energy, and new energy vehicles
Natural gas as % of energy supply	8%
Nuclear as % of energy supply	3%
Hydro, installed capacity	331 GW
Wind, installed capacity	105 GW onshore, 15 GW offshore
Solar, installed capacity	15 GW, including 4 GW rooftop solar PV and 1 GW concentrated solar
Nuclear, installed capacity	40 GW new
Battery Electric Vehicle (BEV) and Plug-in Hybrid Electric Vehicle (PHEV) fleets	500,000 vehicles
New building efficiency	65% energy consumption reduction compared with 1980 building stock

China's Energy and Environmental Landscape

China's electricity capacity was expected to reach 1140 gigawatt (GW) in 2012, surpassing the United States for the first time, with the vast majority of power coming from coal sources (Chen 2012). By comparison, the US electricity capacity was 1,139 GW as of 2010. Coal accounts for 42 percent of the US electricity supply (EIA 2011). Currently, China emits 35 percent more CO_2 per dollar of output than the United States and 100 percent more than the European Union. The increase in energy-related pollution in the past few years in China has been driven primarily by industrial energy use, mainly fueled by coal.

China's rapidly growing economy, already-large population, and massive coal-fueled energy consumption all threaten its future environmental sustainability. China faces many environmental challenges, including water scarcity, exacerbated by water pollution, and releases of toxic substances in the environment. Coal is at the heart of most of China's environmental woes, with major implications for human health. Most of China's air pollution emissions come from the industrial and electricity sectors, although in the cities, transportation is a major contributor as well. Particulate matter from coal is a major air pollutant. Concentrations of PM10 (particles the size of 10 microns or less that are capable of penetrating deep into the lungs) in China's cities are extremely high, ranging from the extreme of Panzhihua's average concentration of 255 to 150 in Beijing, 140 in Chongqing, and 100 in Shanghai. These numbers can be compared with 45 in Los Angeles and 25 in New York City. PM10 can increase the number and severity of asthma attacks, cause or aggravate bronchitis and other lung diseases, and reduce the body's ability to fight infections. Certain people are especially vulnerable to PM10's adverse health effects; they include children, the elderly, exercising adults, and those suffering from asthma or bronchitis. In addition, each year more than four thousand miners die in China's coal mines, mostly in accidents.

Sulfur dioxide emissions, mostly from coal combustion, rose 46 percent between 2000 and 2010 (Lu, Zhang, and Streets 2011). Acid rain affects southeastern China especially, and Hebei Province is most severely impacted, with acid rain accounting for more than 20 percent of crop losses. Hunan and Shandong provinces also experience heavy losses from acid rain. Eighty percent of China's total losses are estimated to be from damage to vegetables.

The total economic costs of China's air pollution are high. According to a 2007 report from China's government and the World Bank, conservative estimates of illness and premature death associated with ambient air pollution in China were equivalent to 3.8 percent of GDP in 2003. Acid rain, caused mainly by sulfur dioxide emissions from coal combustion, was estimated to cost 30 billion RMB in crop damage (mostly to vegetables). Although water pollution is less directly tied to coal consumption, it is still fundamental to human well-being, and it too has become a major drag on overall economic growth. Health damages from water pollution were estimated to account for 0.3–1.9 percent of rural GDP (World Bank, State Environmental Protection Administration of China 2007). A more recent analysis of the health effects of ozone and particulate matter in China determined that these types of pollution led to a 5–14 percent loss of welfare for each year of the study period (1975–2005), and that for 2005, the total welfare loss was US$111 billion in 1997 dollars (Matus et al. 2012).[2] An update published by the World Bank and Development Research Center of the State Council (2012) in China found that the costs of environmental degradation and resource depletion in China approached 10 percent of GDP over the past decade, with air pollution accounting for 6.5 percent, water pollution 2.1 percent, and soil degradation 1.1 percent.For the first time in its history, the Chinese government included a compulsory target to reduce carbon intensity by 17 percent between 2011 and 2015 in its twelfth five-year plan (see table 2.2). This important policy change indicates that the Chinese government has finally decided to make a decisive global contribution to preventing climate change. The twelfth five-year-plan thus represents a decisive break from the past. Not only does it include the new carbon intensity target, but it also has targets for energy intensity, the proportion for nonfossil sources in the overall energy supply, natural gas, renewables, nuclear, and so forth. In addition, it declares that China must set a reasonable limit on the total amount of national energy consumption (not just energy intensity), although it does not provide a specific target. It also states that China should establish a carbon market "step by step" (Xuan and Gallagher 2013).

Environmental Laws and Climate Policies

China has an extensive range of environmental laws, including seven overarching environmental laws, fifteen natural resources laws, twenty-eight environmental administrative regulations, twenty-nine environmental standards, and more than six hundred local environmental rules (Ministry of Environmental Protection 2012; State Council Information Office 2007). The key challenge with environmental laws and regulations in China is in their implementation. Many environmental regulations are top-down in nature, meaning they come from the central government, but their implementation must take place at the local level, where the environmental challenges occur. The relatively weak central government authority that oversees environmental regulation in China has not been successful at encouraging implementation at the local level. The enforcement of environmental regulations is generally less of a priority for local officials than ensuring that economic growth targets are met. In March 2009, the State Environmental Protection Agency was upgraded to the Ministry of Environmental Protection, although it remains to be seen whether this change helps with the challenge of implementation of current laws and regulations. China has some stringent environmental regulations in place, but many of the standards and targets are not being met. Many foreigners assume that because China is a centrally planned economy, the government can easily implement any policy that it wishes to enforce. Yet the reality is that because of China's vast population, huge number of enterprises and factories, and limited environmental monitoring and enforcement capacities, many environmental policies are inadequately enforced.

Coal

China is the largest producer and consumer of coal in the world. Until recently, China was self-sufficient in coal supply, but it gradually began importing coal during the 1990s. In 2009, China became a net importer of coal for the first time, and in just two years, by 2011, it was the largest coal importer in the world. Most of its coal imports are high-quality or precisely blended coals. Because of the coal intensity of China's energy supply, it has become the largest overall emitter of greenhouse gases in the world, though not the largest on a per capita basis or in terms of cumulative historical emissions.

The power sector dominated coal consumption at 49 percent as of 2009, with iron and steel accounting for 9 percent, minerals 8 percent, chemicals 6 percent, and the residential sector down to only 3 percent (Tu 2011). China's energy infrastructure is presently locked into coal consumption, and this structural reality presents a gigantic challenge to its ability to green its economy. Investors and the government are unlikely to be willing to prematurely retire existing infrastructure given how costly and long-lived it is, and so far, higher-priced lower-carbon options are typically not chosen for new projects unless special incentives exist. On the other hand, the Chinese government has shut down many GWs worth of small and inefficient power plants, as discussed below.

Thus, China continues to lock itself into a high-carbon future with each new coal-fired power plant or factory that is built there. The two technological options that can mitigate CO_2 emissions from coal are efficiency and CCS. China is rapidly moving to more efficient coal technologies, and beginning to support research and demonstration projects in CCS as well. The Chinese government has made a major effort to improve coal use efficiency, and as of 2011, the average power plant efficiency had reached 37 percent, which is higher than in the United States. In improving the efficiency of coal use, the government has shut down thousands of small and inefficient coal-fired electricity plants, and replaced them with large, higher-efficiency ones. Indeed, China leads the world in the construction of the most efficient kind of coal-fired power plant, ultra-supercritical coal plants, of which China had thirty-nine (1,000 megawatt [MW]) as of 2011 (Z. Li 2012). China's GreenGen project, an IGCC plus CCS demonstration project led by China Huaneng Group, the largest electricity company in China, and originally inspired by the ill-fated FutureGen project in the United States, began construction in 2009. In its second phase, it will capture CO_2 at a commercial scale.

The Chinese government controlled coal prices until 1993, when a deregulation process began that continued through the 1990s. Now, the government allows coal prices to be determined mainly by the market, within a band set by the government. In fact, coal prices have risen considerably in recent years, which should have the effect of encouraging greater conservation on the part of users. The government still regulates electricity prices, however, and since most electricity is generated from coal, electricity generators have increasingly complained of not being able to raise their prices in line with rising coal prices. Without higher electricity prices, electricity consumers will have little incentive to reduce consumption.

The hard truth is that coal will continue to dominate China's energy mix for decades to come. Coal has consistently maintained a dominant share of the energy supply in China. Coal accounted for 72.2 percent of the total energy in 1980, and only slightly decreased to 70.4 percent in 2009. China's current heavy reliance on coal presents the largest challenge to its quest for a green economy. The good news is that China is aggressively implementing energy-efficiency measures and developing alternatives to coal, especially nuclear and wind power.

Energy Efficiency

A suite of energy-efficiency and industrial restructuring programs drove China's energy intensity down significantly during the eleventh five-year plan (2006–2010). One of the core elements of China's eleventh five-year-plan period was to lower national energy intensity by 20 percent. Several policies and programs contributed to China's ability to save 527 million tons of coal equivalent between 2006 and 2008 (Price et al. 2011). First, China's Top-1000 program helped to cut energy use among China's biggest energy-consuming enterprises.[3] The 10 Key Projects program provided financial support to companies that implement energy-efficient technology. In addition, many inefficient power and industrial plants were targeted for closure. While China built a reported 89.7 GW of new power plant capacity in 2009, it also shut down 26.2 GW of small, inefficient fossil fuel power stations. Thus, one-third of China's new power plant growth in that year was offset by old plant closures—a significant share of the nation's annual capacity additions. National appliance standards along with other policies and programs at the provincial level accounted for the rest of the improvements.

The government has also strengthened local accountability for meeting targets by intensifying oversight and inspection. China's 2007 Energy Conservation Law requires local governments to collect and report energy statistics, and requires companies to measure and record energy use. In addition, each province and provincial-level city is required to help meet the efficiency goal of the current five-year plan. Governors and mayors are held accountable to their targets, and experts from Beijing conduct annual site visits of facilities in each province to assess their progress. The 10 Key Projects program requires selected enterprises to undergo comprehensive energy audits and offers financial rewards based on actual energy saved. The audits follow detailed government monitoring guidelines and must be independently validated. There have also been increases in staffing and funding in the key government agencies that monitor energy statistics and implement energy-efficiency programs. In 2008 alone, China reportedly allocated 14.8 billion RMB (US$2.2 billion) of treasury bonds and the central budget as well as 27 billion RMB (US$3.9 billion) of governmental fiscal support that year to energy-saving projects and emission cuts.

While industrial energy consumption in China is increasing, it has been offset considerably by energy-efficiency improvements. Recent surges in energy consumption by heavy industry in China caused the government to implement measures to discourage growth in energy-intensive industries compared with sectors that are less energy intensive. Beginning in November 2006, the Ministry of Finance increased export taxes on energy-intensive industries. Simultaneously, import tariffs on twenty-six energy and resource products, including coal, petroleum, aluminum, and other mineral resources, were reduced. Whereas the increased export taxes were meant to discourage relocation of energy-intensive industries to China for export markets, the reduced import tariffs were intended to promote the utilization of energy-intensive products produced elsewhere.

Implemented in response to the increasing energy intensity trends experienced during the first half of the decade, China's energy intensity target and the supporting policies described above seemed to be a successful means of reversing the trend. While the country fell just short of meeting its eleventh five-year plan's energy intensity target of 20 percent (the government reported that a 19.1 percent decline was achieved), there is no doubt that much was learned through efforts to improve efficiency nationwide. Many changes were made to how such national targets are enforced at the local level, including the incorporation of compliance with energy intensity targets into the evaluation for local officials. The twelfth five-year plan builds directly on the eleventh five-year plan's energy intensity target and its associated programs, setting a new target to reduce energy intensity by an additional 16 percent by 2015. While this may seem less ambitious than the 20 percent reduction targeted in the eleventh five-year plan, it probably represents a much more substantial challenge. The largest and least efficient enterprises have already undertaken substantial efficiency improvements, leaving smaller, more efficient plants to be targeted in this second round.

Oil and Gas

Natural gas accounts for only 3 percent of Chinese energy supply, and is not commonly used for power generation in China due to limited domestic conventional gas resources and relatively high prices. Instead, China uses natural gas for fertilizer production and residential fuel supply in some cities. Nevertheless, for the first time ever, the Chinese government included a target for the use of natural gas in the twelfth five-year plan, and one of the anticipated sources of natural gas is from China's vast shale gas resources. China has the largest estimated reserves of shale gas in the world at 1,275 trillion cubic feet (*Financial Times* 2012). The twelfth five-year plan includes a target of producing 6.5 billion cubic meters of shale gas by 2015, but China is just beginning to explore shale, which may be constrained in China by the scarcity of water. China also has considerable coal-bed methane resources. As of 2010, China's natural gas consumption had reached 109 billion cubic meters, less than one-sixth that of the United States. China's domestic natural gas production was 96.8 billion cubic meters, much of which came from the Tarim Basin in China's Xinjiang Province. China's largest gas pipeline stretches 2,500 miles from the Tarim Basin to China's eastern provinces. The main foreign sources of China's gas imports were Australia, Indonesia, Malaysia, and Qatar, in that order. LNG corporations accounted for 78 percent of China's natural gas imports (BP 2011).

Oil consumption is primarily driven by the growth in automobiles in China. Beginning in the late 1990s, the automobile industry grew extraordinarily rapidly, and as of 2011, China produced 18.4 million cars in the world, more than any other country (Z. Li 2012). China is now the second-largest oil consumer and oil importer in the world. Its foreign oil dependency was only 40 percent in 2005, but it is growing quickly, and as of 2011, had almost reached 60 percent.

Renewable Energy

China's promotion of renewable energy was kick-started with the passage of the Renewable Energy Law, which became effective on January 1, 2006. The law created a framework for regulating renewable energy and was hailed at the time as a breakthrough in the development of renewable energy in China. It created four mechanisms to promote the growth of China's renewable energy supply: a national renewable energy target, a mandatory connection and purchase policy, a feed-in tariff system for wind energy, and a cost-sharing mechanism, including a special fund for renewable energy development. Several additional regulations were issued to implement the goals established in the Renewable Energy Law, including pricing measures that instituted a surcharge on electricity rates to help pay for the cost of renewable electricity, plus revenue allocation measures to help equalize the costs of generating renewable electricity among provinces.

In addition to the Renewable Energy Law, the 2007 "Medium- and Long-Term Development Plan for Renewable Energy in China" put forth several renewable energy targets, including a nationwide goal to raise the share of renewable energy in total primary energy consumption to 15 percent by 2010 (later revised to refer to all nonfossil sources, including nuclear power). In addition to this ambitious target, the government established a number of specific but complementary policies to boost renewable energy generation.

Power companies have mandatory renewable energy targets for both their generation portfolios and annual electricity production that they must meet. In December 2009, amendments to the Renewable Energy Law were passed, further strengthening the process through which renewable electricity projects are connected to the grid and dispatched efficiently. The amendments also addressed some of the issues related to interprovincial equity in bearing the cost of renewable energy development.

Under the twelfth five-year plan, new policies and targets have been set forth. Initially, this plan targeted 5 GW of installed capacity for solar PV, but it quickly raised this number to 15 GW by 2015 (see table 2.2). It intends to achieve 4 GW of rooftop capacity by 2015 and 1 GW of concentrated solar power. The longer-term targets are 50 GW of solar PV by 2020 and 70 GW of wind power by 2020. In 2011, the government established a solar feed-in tariff that was added to the Golden Sun PV demonstration policy. The government also set a new target for biomass-based power of 13 GW by 2015.

China's Energy Innovation System

The energy innovation system in China is dynamic and rapidly evolving. The system has never been formally characterized, although parts of it have been described or analyzed (see, for example, Grübler et al. 2012; Marigo, Foxon, and Pearson 2008; Zhao and Gallagher 2007). A complete description is not attempted here; rather, an overview of the key actors, magnitudes, and trends in China's energy innovation system is provided.

In a comparative assessment of public investments in energy research, development, and demonstration (RD&D) in the major emerging economies, China stood out as the largest government investor with public investments of approximately US$11.8 billion, plus an additional US$1.3 billion from state and local governments, and partially state-owned enterprises as of 2008. These investments can be compared with US government energy RD&D investments of $4.1 and $2.5 billion, respectively, from state and local governments. In sum, the Chinese government appears to be investing approximately three times as much as the US government in energy RD&D, and these figures do not include support for market formation or deployment. It is also important to note that these figures do not include estimates for Chinese government investments in electricity transmission, distribution, and storage or renewable energy since no data could be found for these categories (Kempener et al. 2010; Gallagher, Anadon, et al. 2011). According to a different estimate by Jiang Kejun for the Global Energy Assessment (Grübler et al. 2012), universities and R&D institutions were funded at the level of 3,415 million RMB (US$54 million) to conduct energy R&D, and industry conducted energy R&D at the level of 23,334 million RMB (US$370 million) in 2004. The total government investments were estimated by Jiang to be 4395 million RMB (US$69 million) and the total enterprise investments were 22,432 million RMB (US$355 million) in 2004. The figures provided by Jiang have many gaps as well, including data on government investments in end use, efficiency, transmission, and distribution as well as government and enterprise investments in renewable energy. The Jiang figures are much lower even than the Kempener and colleagues data for 2004, which reflects the fact that state-owned enterprises are included in the figures from Kempener and his colleagues.

Still, these data are not consistently reported by the Chinese government and should thus merely be considered first-order estimates. Although there are severe data gaps in both sources on efficiency and renewable energy, investments in fossil fuel R&D are emphasized, which may mean that the Chinese energy R&D portfolio is dominated by supply-side investments, which in turn are dominated by fossil-fuel-related research.

The above investment data provide a relatively hazy picture of the Chinese energy innovation system because little analysis has been done to compare the quality of Chinese energy RD&D outputs internationally or even among different sectors in China. Certain energy firms stand out as taking a leadership position on energy innovation in China. Huaneng is the largest investor in GreenGen, the first IGCC plant to be built in China. The electricity company is also conducting pilot-scale experiments with carbon capture, both pre- and postcombustion. Shenhua Group, the largest coal company in China and the world, has created a National Institute of Clean-and-Low-Carbon Energy (NICE) with support from the Ministry of Science and Technology (MOST). The two firms are among the biggest state-owned energy companies in China, and so perhaps it is not so surprising that they are making large and symbolic efforts to do directed RD&D on cleaner energy technologies. Other private firms in China are also well known for their innovation efforts in clean energy, including BYD, the battery producer in which Warren Buffet invested. In an interview, a senior BYD official claimed that BYD had invested 4 billion RMB (US$633 million) into electric vehicles (EV) (interview 21, 2010). These illustrations are encouraging anecdotes, but they do not reflect a comprehensive analysis of the strengths and weaknesses of enterprise-level energy RD&D in China.

At the central government level, there are three main entities that focus on energy RD&D: MOST, the Chinese Academy of Sciences (CAS), and the National Natural Science Foundation (NNSF). Within MOST, there is an energy division that is responsible for all energy activities and funding. There are also special projects offices within MOST, many of which have energy-related projects under their control. The energy division has a number of expert committees that advise it on its technology strategy as well as help to decide where to allocate funding. There are expert committees on gas turbines, clean vehicles, and coal gasification, for example (many of the people interviewed for this book are or have been members of these committees). The total 2011 budget for MOST is 24.69 billion RMB (US$3.8 billion), including 24.1 billion RMB on science and technology research funding (Jia 2011). Within CAS, there are numerous institutes that focus on cleaner energy technology including the Wuhan Institute for Soil and Rock Mechanics, Institute for Thermoengineering Physics, Guangzhou Institute of Energy Conversion, and Research Center for Clean Energy and Power, among many others. CAS has a separate budget from that of MOST. The total R&D expenditure for CAS in 2010 was 233 billion RMB (US$37 billion), but no subfigures for energy are available. The NNSF is the principal funding agency for science in China, and its budget in 2010 was 10.4 billion RMB (US$1.6 billion). It is important to mention that the National Development and Reform Commission (NDRC) along with its related Energy Bureau are critical to both demonstration and early deployment of cleaner energy technologies because the NDRC must approve all major construction projects. Many other agencies affect the energy innovation endeavor in China, including the Ministry of Education, Ministry of Environmental Protection, Ministry of Agriculture, and Ministry of Finance. It is also crucial to recognize that there are similar agencies at the provincial level, and that the local governments provide a great deal of funding for both R&D as well as demonstration projects for clean energy.

All the ministries and institutions mentioned above—central, provincial, and local as well as state-owned enterprises—are guided by the five-year plans approved by the National People's Congress. Longer-term plans are issued periodically, such as the fifteen-year "Medium- and Long-Term Plan for the Development of Science and Technology" issued in 2006, and the 2007 "Medium- and Long-Term Development Plan for Renewable Energy in China." The energy-related targets of the twelfth five-year plan have already been discussed earlier in this chapter, but the science and technology plan has not. This document was the origin of China's controversial "indigenous innovation" policies because it commits China to develop them in twenty strategic industries, including energy resources, recycled economy (improved efficiency in the use of resources), and transportation. The plan calls for China to become one of the top five countries in the world in the number of patents granted to Chinese citizens, increase expenditures on science and technology as a percentage of GDP, and reduce its growing reliance on foreign technology (Cao, Suttmeier, and Simon 2006).

Conclusion

As the largest energy-consuming country and largest greenhouse gas emitter on the planet, China is an essential laboratory for the study of the global diffusion of cleaner and more efficient energy technologies. The energy paradox of China is that while it is the largest producer and consumer of coal in the world, it is simultaneously the largest investor in renewable energy technologies and is reducing the energy intensity of its economy faster than any other country today. The Chinese central government has recognized that China must pursue a less resource-intensive mode of economic and social development, and appears firmly committed to making numerous cleaner energy industries a central pillar of its future development. The Chinese government is committing huge sums of capital to the development and deployment of cleaner, more efficient energy technologies. It has planned to prioritize "new energy" for the long term through its medium- and long-term plan for scientific development as well as the eleventh and twelfth five-year plans. It is also taking advantage of its now-unrivaled manufacturing capacity, which is flexible and nimble due to the clustering of manufacturing facilities, a skilled workforce, and a strong underlying infrastructure.

China is trying to apply to the clean energy sector its tried-and-true development model of importing advanced technologies from abroad, localizing them, and then producing them for both the home and foreign markets.[4] As such, it is a fascinating and important place to study international technology transfer in the clean energy sector. In the next chapter, "Four Telling Tales," case studies of technology transfer in four different clean energy industries in China are chronicled.

3

Four Telling Tales

This chapter provides four telling tales about international technology transfer in cleaner and more efficient energy technologies. These case studies supply an empirical basis to answer questions about the barriers and incentives to the global diffusion of cleaner energy technologies as well as test relevant theories about international technology transfer. They offer complete accounts of each case. The cases will be drawn and elaborated on in subsequent chapters as they relate to particular concepts and findings, but the overview is provided here to give proper context.

The Chinese energy sector offered many possibilities for case study analysis because China is acquiring and exporting many different types of energy technologies. The four that were chosen for this study are, as mentioned earlier, gas turbines, advanced batteries for vehicles, solar PV, and coal gasification. At the outset of the research process, these cases appeared heterogeneous in terms of Chinese technological capabilities, manufacturing skills, penetration in the Chinese market, and extent of exports. No substantial case studies on international technology transfer regarding these technologies in the Chinese context existed, although happily some prior research had been done on some aspects of their technological development in China or elsewhere, for example, for solar PV (Liu et al. 2010; de la Tour, Glachant, and Ménière 2011; Fu and Zhang 2011; Zhao et al. 2011), advanced automotive batteries (Harwitt 1995; Gallagher 2006a; Nam 2011), gas turbines (Auer 1993; Watson 2004; Smil 2010), and coal gasification (Watson and Oldham 2000; Zhao et al. 2008; Cai et al. 2009; Chen and Xu 2010; Liu and Gallagher 2010). Substantial research already existed on wind development in China (Lewis 2007; Ru et al. 2012; Lema and Ruby 2007; Liu and Kokko 2010; Li, Shi, and Hu 2010). Gas turbines seemed to be a case of the relative "failure" of international technology transfer, and solar

PV a case of the relative "success," with coal gasification and advanced batteries falling somewhere in between. Relatively little technology transfer had occurred in the case of gas turbines, whereas substantial technology transfer had taken place in terms of solar PV.

The main barriers and incentives to international technology transfer in these four cases are analyzed toward the end of this chapter. The perspectives of Chinese interviewees are contrasted with the perspectives of foreign interviewees, and areas of consensus and divergence are identified. To preview the conclusions, there is widespread agreement that the most important incentive is national-level, market-formation policies. Access to low-cost capital is also a crucial incentive, or conversely lack of access to affordable finance presents a major challenge to firms. There is clear agreement among Chinese and foreign firms that the intellectual property rights environment is already strong or improving in China, but overall, IP was not found to be a major incentive or barrier, despite the fact that foreign firms all express concern about IP protection. Smart business practices on both sides of the technology transfer transaction are found to greatly facilitate the process. Chinese absorptive capacities are generally strong, and this also facilitates the technology transfer process. Everyone agrees that governments must correct for market distortions to better enable cleaner technologies to compete against incumbents. The marketplace does not typically value the benefits of reducing greenhouse gas emissions, improving energy security, protecting public health, and other indirect advantages. Failure to correct for these externalities is the biggest barrier to clean energy technology transfer. Non-Chinese firms have problems accessing capital at home for expansionary or export activities, but Chinese firms do not have this problem. Lack of international experience and/or a global perspective on markets can be a significant barrier to the acquisition and sales of technology.

Gas Turbines

The tortured history of gas turbine development in China presents the best evidence of barriers to the international transfer of cleaner energy technologies of any discussed in this book and perhaps in the world. Export controls on the part of foreign governments, weak and inconsistent R&D and industrial policy on the part of the Chinese government, poor Chinese firm-level capabilities, high costs, and an oligopoly of foreign firms have all conspired to limit transfer of gas turbine technology to China. According to one Chinese expert, "Lack of gas turbine technology may constrain the diffusion of clean energy technologies [in China] more than anything else."[1]

The Chinese electricity industry could use gas turbine technology to help meet peak loads, provide backup for renewable energy sources, and enable switching from coal to reduce CO_2. The advantages of natural gas combined cycle technology (NGCC) are many, including its high degree of energy efficiency, flexibility in operation, and short construction time. Gas turbines can be used for power generation as well as industrial purposes. Heavy-duty turbines grew out of competencies that firms developed during World War II for aircraft jet engines that could be applied for use in the electricity industry. Both GE and Siemens were involved in jet engine development programs during and after World War II, and they were the first to develop the gas turbine as a new electricity-generating technology (Watson 2004). Many countries, perhaps most notably the United States and United Kingdom, have been transitioning to NGCC technologies for economic and environmental reasons for the last couple of decades; indeed, NGCC is one of the most efficient and least polluting fossil fuel power-generation technology options available today. There are only a handful of NGCC plants in China, largely due to the high price of natural gas there. Gas turbines are also critical to the coal-fueled IGCC plants because a gas turbine is needed to utilize the synthetic gas generated during the gasification process. IGCC technology is of particular interest because it is relatively easy to separate out CO_2 during the gasification process, and then capture it for reuse in industrial applications or geologic storage.

A key attribute of natural gas turbines is the efficiency of the turbine itself, and the more advanced the turbine technology, the greater the efficiency. As Siemens states in its brochure for its new H-class turbine, "As fuel is the largest single cost item for running a power plant, an increase of 2 percentage points can save the operator millions of Euros over the entire lifecycle of a combined cycle plant with a capacity of 530 MW" (Ratliff, Garbett, and Fischer 2007). Due to this fact, Chinese buyers are keen to obtain the most efficient turbine available not only for NGCC applications but also for IGCC applications since the capital cost of this type of plant is already high, and these plants suffer an efficiency penalty once CO_2 is captured.

The relatively high cost of natural gas in China has presented a persistent hurdle to the use of natural gas in the electricity sector. Because gas supply has historically been limited and expensive, it was mainly used for fertilizer production and residential purposes. Gas was even further disadvantaged because the Chinese government suppressed coal prices until 1994 (Wang 2007). The most advanced turbine price is still a hurdle even for Chinese firms with government backing. Chinese interviewees emphasized the importance of being able to purchase highly efficient turbines at a reasonable cost, and their perception is that the most efficient turbines are so expensive that China has no choice but to develop its own technology. As one expert asserted, "China cannot suffer these extremely high costs, so China must develop its own technology."[2]

There are so few leading manufacturers of gas turbine technology around the world that the industry is oligopolistic by nature. While there is no evidence of cartel-like behavior today, for the first sixty years of its existence, the power plant equipment industry operated through a series of cartels (Watson 2004). As exemplified by the quote above, from the point of view of developing countries, the worries are that an oligopoly of turbine suppliers may set the price for consumers, and also that barriers to entry grow the longer the oligopoly exists due to economies of scale and patent holdings. A third worry is that the oligopoly could refuse to sell desired technology to certain customers, but at least one of the leading firms, Siemens, stated on the record that it was willing to sell its most advanced turbine to China as a final product, indicating there is *access* to the technology even if the price is high. The slowness of China to begin using gas can be attributed to the historical dominance of coal in China's energy reserves. With vast coal reserves and relatively small conventional gas reserves, it is not surprising that the new Communist government after World War II emphasized the use of coal. For decades, the Chinese government concentrated efforts on coal and oil production, and effectively forgot about gas. One China expert commented that historically, "the production of oil distracted from gas development."[3] Indeed, China's first major gas pipeline, the west-to-east pipeline, was not completed until 2004. Ironically, it now appears that China has the largest shale gas resources in the world, with 1,275 trillion cubic feet of technically recoverable shale gas resources, which is roughly triple the US estimate of 482 trillion ft^3 and more than ten times the estimate of China's conventional gas reserves (Pfeifer 2012, 7).

Chinese scientists are critical of the government's historical innovation strategy and support for the development of gas turbines, characterizing the effort as "fickle" and at a "nursery school" level.[4] As the west-to-east pipeline was under construction, the Ministry of Science and Technology began planning an RD&D program for gas turbines under the high-tech 863 program. Meanwhile, the NDRC coordinated the effort to transfer gas turbines to China, setting conditions that if foreign firms wanted to sell their technology, they needed to find partners in China, as seen in table 3.1 below. Dozens of E- and F-class gas turbines were imported between 2003 and 2006. Typically, R&D funding from the Chinese government is a carrot that leads and guides Chinese firms in certain directions, but funding for gas turbine research was limited. The Ministry of Science and Technology could only provide 7–20 percent of the cost during the tenth and eleventh five-year plans, and it was often the case that the Chinese firms could not pay the remaining share.[5] Another problem was that research funding for gas turbines was historically divided into different ministries that are notoriously bad at coordination, so the funding has not always been invested wisely and effectively.[6]

Chinese scientists are equally critical of China's major firms for failing to invest in and develop better technology. Most of these firms are state owned, and their leadership is determined by the part of the government that owns them—almost akin to a "political appointment" in the United States. As one scientist commented, "No leader is responsible for 10–20 years, they are only there for 1–2 years."[7] In other words, the CEO of a Chinese state-owned enterprise does not have personal incentive to make long-term investments for the good of the firm but rather to maximize profit in the short term. Due to state-owned enterprise reforms, the major Chinese firms had tiny profits and therefore lacked funds to invest prior to the 2000s. Once a real market developed for advanced electricity-generation equipment, the Chinese firms were not prepared to serve it with their own technology. Each firm had no choice but to buy technology from foreign firms if they wanted to remain relevant in the market, and they spent all their money importing technology, which further crowded out indigenous RD&D.

For all their pessimism, doom, and gloom, the Chinese scientists and engineers are highly esteemed by their foreign counterparts. All the foreign experts interviewed believe it is only a matter of time before the Chinese catch up in gas turbine technology. "Global technology leadership is what is at risk," according to a company official in one foreign firm.[8] In chapter 5, patenting trends are examined, and we will see that gas turbine technology is the only clean technology where patents granted to Chinese applicants still considerably lag foreign patents, although they are indeed growing. This finding lends credence to both perspectives: Chinese scientists and firms are beginning a catch-up process, but are still far behind in terms of their technological capabilities.

Table 3.1

Foreign–Chinese firms gas turbine relationships

Chinese Firm	Foreign Firm	Type
Shanghai Electric (49%)	Siemens (51%)	Siemens Gas Turbine Parts Ltd. joint venture for high-temperature parts (2005–present) and assembly of final product
Shanghai Electric	Siemens	Framework purchase and sales agreement for power generation, transmission, and distribution equipment (2009–present)
Dongfang Turbine Company (49%)	Mitsubishi (51%)	MHI Dongfeng Gas Turbine Co. joint venture for gas turbine core components for 701 gas turbine, marketing, and after-sales service (2004–present)
Dongfang Turbine Company	Mitsubishi	License agreement (2003–present)
Harbin Electric	GE	Different purchase agreements over time
Nanjing Turbine and Electric Machinery	GE	Joint venture for E-class assembly with some locally manufactured parts and joint bidding (125 MW) (2004–present)
Huadian Group (51%)	GE (49%)	Huadian GE Aero Gas Turbine Equipment Co. Ltd. joint venture to produce aero-derivative gas turbines for distributed generation
Shenyang Liming Aero-Engine Company, Ltd. (49%)	GE (51%)	GE–Liming joint venture (2003–?) to produce components to contribute to assembly of E- and F-class gas turbines

Sources: Nanjing Turbine and Electric, http://www.ntcchina.com/website/en/Company/frmCompanySur.aspx; Siemens, press release, http://www.financial express.com/news/siemens-signs-gas-turbine/49156; http://www.globalsecurity. org/military/world/china/liming.htm; interviews 7, 76, 51, 66, 41, 78; MHI, press releases, June 15, 2010, and March 7, 2003.

Internationally, the three major producers of heavy-duty natural gas turbines for power generation are Siemens, GE (which acquired Alstom's heavy-duty gas business in 1999), and Mitsubishi, and each of them has a relationship with a Chinese counterpart (see table 3.1). All three big firms have formed joint ventures to assemble older gas turbine technology. Purchase and license agreements are also prevalent. Siemens has a joint venture with Shanghai Electric to produce parts for gas turbines as well as a framework purchase and sales agreement with Shanghai Electric for power generation, transmission, and distribution equipment. In 2011, Shanghai Electric bought 1,247 million RMB worth of products from Siemens, and Siemens bought 135 million RMB worth of products from Shanghai Electric. GE has also used a purchase agreement model with Harbin Electric, and in 2011, Harbin agreed to purchase four of GE's F-class gas turbines. Mitsubishi has a joint venture with Dongfang Turbine Company for gas turbine components as well as marketing and after-sales service.

Government export controls were raised by both American and Chinese interviewees as absolute barriers to the transfer of gas turbine technology from the United States to China. The US government imposes these controls through Export Administration regulations because of the potential "dual use" of technologies for both civilian and military applications. In order to export items, US firms must obtain a license from the US Department of Commerce for export. In a 2010 review of US export controls on "green technology items," the US Bureau of Industry and Security confirms that machine tools for the production of industrial gas turbine components, the components themselves, the design technology, the technology for reducing emissions from gas turbines, and the encryption for the control systems would all be subject to a license requirement to export to China and many other countries. Other cleaner energy technologies related to solar, wind, alternative fuel vehicles, LEDs, and other energy-efficiency technologies are also subject to export controls. On the whole, however, the bureau found that "most green technology-related items do not require a BIS export license," and that of the $1.3 trillion in total US exports in 2008, $75 billion was green technology, and of that, only 0.9 percent ($694 million) required an export license. The report does not disclose how many applications for a license were rejected or deterred (Watts and Bagin 2010, 1).

Because gas turbines epitomize high-end, advanced technology for the leading three firms, and because these firms have invested millions of dollars in their development, they are concerned about protecting their intellectual property. Steps these firms have taken to protect their IP include only licensing older technology, creating joint ventures with Chinese firms to produce certain components, and then shipping others to China for final assembly. One firm interviewed maintained it was not willing to sell the most advanced gas turbines to Chinese buyers at any price, even as a final product.[9] On the other hand, Siemens company officials insisted they would be willing to sell their latest H-class turbine, which they reportedly spent $500 million euros to develop (Weiss 2012). Dr. Hans-Peter Böhm of Siemens states, "We are not willing to make a technology transfer of the H-class turbine, but we are ready to sell the ready-made product" (personal communication).

Advanced Batteries for Cleaner Vehicles

China made a major push toward EVs beginning in its eleventh five-year plan. Given China's heavy reliance on coal for electricity supply, the main environmental benefits of this shift to EVs could be cleaner air in some cities and a reduction in noise pollution. According to a recent analysis by Shuguang Ji and his colleagues (2012), however, replacing gasoline cars with EVs in China with its current electricity supply mix will result in *higher* CO_2 emissions and increased mortality risk from $PM_{2.5}$ in most Chinese cities. The Chinese government nonetheless views a shift to EVs to be beneficial to China's energy security. China has already become a net importer of coal, and its current reserve-to-production ratio of coal is only thirty-three years (BP 2012). The energy security benefits of EVs fueled from coal-based electricity are therefore somewhat dubious.

This inquiry on advanced batteries for vehicles was originally intended to focus on the development and deployment of lithium-ion (li-ion) batteries for HEVs, EVs, or plug-in hybrid-electric vehicles (PHEVs). As it turns out, Chinese technological capabilities are still weak in li-ion batteries, and sales of them for EVs, HEVs, and PHEVs in China are negligible. Chinese firms manufacture li-ion batteries for automobiles using imported technology largely for export. In doing so, they draw on their manufacturing skills developed for battery applications for consumer electronics. The Chinese government recently affirmed its determination to support RD&D and market-formation activities for EVs and PHEVs in China.

Historically, the Chinese government strategy for cleaner automotive technology has not been clear or stable. When the government finally developed a definitive industrial policy for the automobile industry in the 1990s, little thought was devoted to fuel-efficiency or pollution-control technology. China's first pollution-control standards for automobiles did not take effect until 2000, and its first fuel-efficiency standards did not take effect until 2005 (Gallagher 2006a; Oliver et al. 2009). Still, during the late 1990s, MOST was beginning to make significant investments in cleaner vehicle technologies, and EV research was included in MOST's 863 "high-tech" research program starting in 2001. At that time, there was still indecision about the appropriate technology pathway, with some of the Chinese experts including Wan Gang (who went on to become the minister of science and technology) convinced of the merits of fuel cell vehicles, and others pushing for an electrification pathway. At that time, the United States was also making a strong bet on hydrogen fuel cell technology with the FreedomCar RD&D program.

Even though MOST's interest in and commitment to cleaner vehicle technologies, and batteries specifically, has not waned, the R&D program budget has never been large. At the beginning of the eleventh five-year plan MOST invested 1.1 billion RMB (US$174.5 million) in a new technology road map for the EV industry and its overall R&D program. In 2010, the 863 Key Technology and System Integration Project for EVs announced 738 million RMB (US$117 million) for battery and EV integration with 42 percent of the funds to be devoted to battery research (Earley et al. 2011). It's not clear whether or not these investments duplicate each other, or whether these are separate funds. In addition, local governments and enterprises invested 7.2 billion RMB during the eleventh five-year plan (Wang 2013). For the twelfth five-year plan, MOST allocated 1.2 billion RMB for EV R&D (ibid.).

MOST invested in both fuel cell and electrification approaches, and even established an EV demonstration project as early as 1997 in Shantou, Guangzhou, which I visited in September 2001. At that time, engineers there were monitoring and testing about twenty-five vehicles. Five of them were donated by Toyota, five by GM, and the rest were Chinese, mostly minibuses and trucks. I test-drove many of the models available. Taxi drivers were using some of the vehicles in the city, but the slow charging time was not compatible with their schedule because the cars were used day and night (by different drivers), so there was insufficient time for charging. Even at that early date, the manager of the demonstration project commented that the biggest barriers to the further development of EVs were "coordination among agencies, national and local" as well as a comprehensive national policy on EVs. Other barriers cited at the time were the costs of the batteries, their durability, and the need for a national certification and testing center.[10] As of 2012, together with three other government agencies, MOST is supporting testing of thirteen thousand all-electric and alternative energy vehicles in twenty-five cities through a Ten Cities, Thousand Vehicles program, mostly in the form of taxis and buses used for public transportation (Earley et al. 2011; McDonald 2012).

After the turn of the twentieth century, the debate about the correct technological pathway continued, with the Chinese government now leaning toward an electrification pathway. According to interviews, government officials now see several advantages in leapfrogging directly to pure EVs. First, they won't have to technologically master the internal combustion engine—they can avoid it entirely. After many years of producing automobiles through joint ventures with foreign firms, they feel that they have learned little about the key technologies like fuel injection, motors, and transmissions besides how to manufacture them well. For EVs, the argument goes, there are really only two core technologies—the battery and the electric engine—so by choosing an electrification approach, the Chinese believe they can avoid having to master the internal combustion engine and have a better chance of developing indigenous technology. Although hybrid-electric technology is attractive, it uses an internal combustion engine, and there is the further problem of the Japanese patents. The Chinese have not been able to obtain licenses from Toyota for hybrid-electric technology, and they further think there is no room for Chinese innovation because the Japanese firms have defensively patented the entire space. Third, pure EVs most decisively resolve the energy security threat so long as domestic fuels are used.[11]

Initial targets were set in 2009 for 500,000 all-electric cars by 2015 and 5 million by 2020, and these goals were reiterated in April 2012 in a new "Energy Saving and New Energy Automotive Industry Plan" (2012–2020). The market-formation policies in support of meeting these targets are evolving. At the end of 2011, the government announced it would invest US$1.5 billion per year to develop and deploy clean vehicle technologies. The government also plans to invest in a charging infrastructure and procurement as well as develop a plan for recycling batteries. Plug-in hybrids will qualify for some of these subsidies (Bloomberg News 2012). The central government in China now offers a 60,000 RMB purchase subsidy to consumers who buy EVs. In addition, starting in 2013, the Chinese government is allowing six cities to experiment with subsidies to individual consumers who purchase EVs. These cities will add 40,000–60,000 RMB in purchase subsidies. As of March 2013, there were approximately 28,000 EVs registered in China, of which about 80 percent were public buses, not including electric bicycles, which numbered 135 million as of 2010 (Jie and Hagiwara 2013).

Chinese interviewees all expressed frustration with the government's failure to resolve the charging infrastructure problem as well as the lack of attention to improving customer acceptance of the technology. As one interviewee commented, "We should strengthen propaganda."[12] Core to achieving all of these targets from the Chinese perspective is the development of their own, lower-cost advanced technologies so that the Chinese do not need to pay high licensing fees to use foreign technologies, or worry about infringing on foreign patents. One Chinese government official noted that competition among foreign firms has caused license fees to come down. But according to him, license fees are generally a little higher for cleaner energy technologies than for conventional alternatives.[13]

The main barriers to the deployment of advanced batteries in China are their cost, the historically inconsistent policies in China, lack of attention to infrastructure, and to a lesser extent, access to foreign intellectual property (various interviews; Wang and Kimble 2011). All the experts interviewed identified cost as the number one barrier to the greater global deployment of advanced-battery technologies. One of China's most well-regarded experts on EV technology remarked, "You know I have been studying the EV situation for many years now, and in China, the most important barrier is price."[14] According to him, market research concluded that if a consumer's initial investment in an EV can be paid back in two years or less, then the consumer will buy the car. A big reason why the advanced vehicles cost more is because of the price of the batteries. Some firms acknowledged that for the domestic market, they sometimes just use lead-acid batteries because they are so much cheaper. As one interviewee explained, li-ion batteries are five to ten times the cost of lead-acid batteries.[15] Chinese firms manufacture li-ion batteries for export and have not had trouble importing the required technology to manufacture these advanced batteries.

According to the Chinese, importing foreign HEV technology is prohibitively high, not only because of the cost of the batteries (which are nickel-metal hybrid batteries, not li-ion), but also because of the cost of importing all the parts and components, which is compounded by Chinese import tariffs (Gallagher 2006c). As of 2010, the Prius was selling for about US$50,000 (280,000–320,000 RMB), about twice what it costs in the US market. One source noted that the Prius was selling for 229,800–279,800 RMB in 2012. Even though Toyota now manufactures the Prius in China in cooperation with First Auto Works, it brings all the parts from Japan (Ockwell et al. 2008). For the batteries themselves, one estimate was that the cost for a li-ion battery produced in China was approximately 1.6 RMB/watt-hour versus 0.6 RMB/watt-hour for a conventional lead-acid battery. Interviewees estimated that with economies of scale, the li-ion battery costs would decline by 25 percent, assuming an annual production of ten thousand.[16]

China has been accused of favoring domestic EV firms through "indigenous innovation policies," but during the twenty-second session of the Joint Commission on Commerce and Trade, in 2011, China confirmed that "it does not and will not require foreign automakers to transfer technology to Chinese enterprises nor to establish Chinese brands in order to invest and sell in China's fast-growing market. China also confirmed that foreign-invested enterprises are eligible on an equal basis for electric vehicle subsidies and other incentive programs for electric vehicles" (US Trade Representative 2011). Foreign battery firms had previously complained that only domestic firms qualified for Chinese subsidies, which led them to conclude that their only choice was to form joint ventures with Chinese firms where they have a minority stake. On the other hand, they see China as a rapidly growing market with policies that are relatively stable, in contrast with the United States. They also note China's longer-term commitment, which also contrasts with the US government. Furthermore, they explain that for nonautomotive applications, many of their clients are in China, so it makes sense to produce and deliver within the same market.[17]

The three major li-ion Chinese battery manufacturers in China are Lishen, BAK, and BYD. Of these three, BYD is the only one that is privately owned, and as is often mentioned, BYD received investment from Warren Buffet. BYD is known for investing much more heavily in R&D, but also for not producing high-quality products, in part due to its heavy reliance on cheap labor.[18] BYD claims it has invested four billion RMB into EV technology R&D. It also formed a cooperative arrangement with Daimler to do R&D for EVs. BYD asserts that its technology acquisition strategy was to buy foreign firms as necessary.[19] All the firms have hired foreign consultants to help with the manufacturing process, mainly from Korea and Japan, but also from Europe.[20] The Chinese li-ion battery manufacturers import the core materials from foreign sources, package them, and then sell the final product. Much of the IP is embedded in the materials imported from abroad, so the Chinese have not been able to "learn by doing" through manufacturing. In this sense, the battery industry is similar to the chemical industry because the core technology is in the powder chemistry of the battery. The foreign firms' decision to ship this powder rather than manufacture it in China despite the shipping costs they have to incur reveals the sanctity of the core technology. All the Chinese manufacturers indicated that their labor low-cost advantage is not going to last long because in order to achieve high-quality products, they would need to move to automated production lines. Also, all the Chinese manufacturers noted that the scale-up from li-ion batteries for electronics to automobiles was extremely difficult. One commented, "It's like stirring a bottle. When you stir in a small bottle, it's easy; but if you stir in a large bottle, it's hard!"[21] The main technical challenges, not unique to Chinese firms, relate to power efficiency and heat dissipation.

Chinese industrial policy for advanced vehicles used to be to try to exchange the market for technology, but Chinese experts believe this policy failed. The main capabilities they've acquired are quality management, marketing, and manufacturing (Wang and Kimble 2011).[22] While all these capabilities are good for job creation and growth within China, the lack of design capability makes China vulnerable in several ways. First, dependence on foreign clean vehicle technology makes China more vulnerable in terms of energy security, because if firms or governments become no longer willing to sell to China, China's demand for oil would increase. Second, China will be limited to its own domestic market unless it can meet rising fuel-efficiency and greenhouse gas performance standards abroad. Third, foreign firms have little incentive to cooperate with Chinese firms. One example cited was Toyota's decision to license its HEV technology to Ford Motor Company in the United States. According to the Chinese perspective, Toyota knew that within a few years, Ford would have developed the same technology, so they took the limited opportunity to make some money. The narrowness of this technological gap explains why Toyota was willing to license to Ford and not a Chinese company. In other words, China needs to improve its technological capabilities for its own sake as well as to improve its bargaining power with foreign firms.[23]

Market-formation policy for advanced li-ion batteries for automobiles in China is indirect and comes in two forms. The first type of market-formation policy is the subsidization of EVs, HEVs, and PHEVs through rebates to consumers. The second type of policy is the fuel-efficiency performance standard, which mandates minimum levels of fuel efficiency.

The current Chinese fuel-efficiency standard, while quite progressive and more stringent than the US standard for automobiles (42.2 miles per gallon by 2015 for China compared with 35.5 miles per gallon by 2016 for the United States), does not require the use of advanced batteries to achieve the standard. The standard can be met through use of smaller engines and incorporation of conventional technology. The Chinese government also created fiscal policies based on engine displacement so that higher excise taxes are placed on more inefficient vehicles, but again, these taxes are not aimed at encouraging deployment of the most advanced vehicles, including advanced batteries. The lack of a gasoline fuel tax, differential purchase and ownership taxes, and more stringent fuel-efficiency or carbon-emission performance standards all hinder the more rapid deployment of advanced-battery technology in China (Gallagher 2006a).

Regarding access to finance, firms revealed mixed experiences. Two firms were limited by what their shareholders were willing to do, and two others indicated that while they didn't have problems accessing finance per se, they were unable to expand operations due to management decisions. In fact, one firm claimed they had as many as ten potential investors. The reason for the shareholder reluctance was not disclosed, but an obvious explanation would be the lack of apparent demand from consumers for EVs in China. Another firm asserted that it had absolutely no finance limitations.

Foreigners and Chinese alike identified intellectual property issues as challenges for their businesses. Some foreign firms reported problems with rogue employees working in their operations in China, but none of these problems were close to life threatening for the firms. No court cases related to IP infringement in battery technology were identified during the course of the research. As discussed above, Chinese firms have not been able to persuade Toyota to license any HEV technology and generally find the HEV patent space to be crowded. All the firms interviewed, Chinese and foreign, had applied for many dozens of patents in China, and this rate of applications is consistent with the figures in chapter 5.

Finally, there are two issues related to the development of EVs in China that should be mentioned: whether or not they reduce greenhouse gas emissions, and their resulting demand for rare earth minerals. On the first question, the answer is, of course, that it depends on China's future electricity supply. A recent analysis estimated that life cycle EV emissions would actually be higher than a conventional internal combustion engine if EV cars were charged from two of China's grids, which are highly dependent on coal, and slightly lower if charged from five of the other regional grids. Given the Chinese future targets for nuclear, renewable, and nonfossil electricity supply, the greenhouse gas benefits of EVs could become more pronounced in the future (Earley et al. 2011). A separate analysis concluded that both CO_2 and PM10 emissions with an EV fleet would be higher than gasoline-fueled vehicles, especially if they were hybrid vehicles, given the existing electricity mix (Ji et al. 2012).

Certain rare earth minerals are important for HEVs and EVs. China is abundant in these minerals, but as world demand rises for electrified cars, the price of these minerals could rise if production does not match demand or China decides to halt exports of rare earths as it did in September 2010. According to a US Department of Energy study in 2010, five rare earth elements (dysprosium, terbium, europium, neodymium, and yttrium) pose critical supply risks and/or their loss would strongly affect the supply of cleaner energy technologies. Tellurium, cerium, and lanthanum are designated as "near critical." Neodymium, cerium, and lanthanum are all deemed especially important for EVs (DOE 2010).

Solar PV

The Chinese solar PV module manufacturing industry adopted a highly globalized perspective to source and export technology. The solar PV manufacturing industry is the youngest of the four reviewed in this book, and the explosive growth starting in the 2000s came as a surprise to many observers, including international competitors. Many of the Chinese firms that grew so rapidly during the 2000s experienced an equally precipitous fall beginning in 2012, but this book is not about that story; rather, it about how the Chinese firms acquired and then exported their PV technologies.

The solar industry in China as we know it today is vastly different from the one that existed in the 1980s and 1990s. At that time, solar PV modules produced there were extremely small scale and mainly produced for rural electrification purposes. The PV manufacturing industry was born in the 1970s when three semiconductor factories were transformed into PV factories. By 1995, China had six manufacturers, all of which imported solar cell production lines from the United States, Canada, and other countries, and the total production capacity then was five MW per year. In a survey of the industry during the 1990s, high production costs, capital shortages, insufficient R&D funding, and lack of market-formation support from the government were cited as major barriers to the further development of the industry. At that time, Chinese banks were reluctant to lend to new PV manufacturers for start-up and expansion activities. Huamei Solar Equipment Company in Qinhuang-dao, for example, had to give up an order of four million RMB because of an inability to secure a loan. There were no supportive policies for solar PV deployment until the Brightness Program, and indeed, the solar PV industry had to contend with taxes for importing raw materials, tariffs for importing solar PV systems, value-added taxes that were the highest rate possible at the time, and corporate income taxes. Interestingly, only one of the early production lines still existed as of 2008 (Marigo, Foxon, and Pearson 2008).

Early deployment of solar PV in China was focused on rural electrification. Small-scale Solar Home System projects were government sponsored with international aid, largely through the Global Environment Facility. The first significant central government support for solar PV came through the Brightness Program, which was established in 1996 by the State Development and Planning Commission with an objective of providing renewable power to twenty million rural Chinese without access to electricity by the year 2010 (Yang et al. 2003). In 2002, the Township Electrification Program was launched. Other rural electrification policies and programs were created, including the Renewable Energy Development Project, which provided a direct subsidy, supported by the Global Environment Facility, to Chinese PV companies to help them sell and maintain small-scale distributed PV modules. The PV Renewable Energy Development Project ultimately supported the installation of 350,000 Solar Home Systems, and also helped to upgrade Chinese small-scale PV producer capabilities by the creation of national standards, certification systems, upgrading of testing facilities, and information gathering and sharing activities (Marigo, Foxon, and Pearson 2008).

Many of the leaders within the Chinese PV industry were educated outside China, worked outside China, and returned to found or join new firms. One source claims that 61 percent of the board members of the three largest Chinese PV firms had studied or worked abroad, noting that the CEO of Yingli studied abroad, as did six people in the management team at Trina Solar (de la Tour, Glachant, and Ménière 2011). The CTO at Yingli and CEO of Suntech both received doctorates at the University of New South Wales in Australia. The CEO and vice president of JA Solar both trained in the United States (one in technology, and the other in business). With their sophisticated knowledge about technology, these leaders were able to make smart and careful choices about which technologies to buy from abroad, and what kind of equipment they needed for their manufacturing plants. Shi Zhengrong, the former CEO of Suntech, commented in an interview that one of his strengths was that he "can understand the technology," and therefore knows what he needs to go out and buy. He said, "Buying is a great strategy for acquiring technology." Suntech licensed technology, purchased manufacturing equipment, and bought foreign firms outright. That being said, Dr. Shi in an interview asserted that technologically, the company's main strategy was "joint development," where it works with other universities or technology partners to develop technological solutions (personal communication). Perhaps because they were educated abroad, these leaders also had a remarkably sharp global perspective on where the markets were for solar PV modules. The traditional Chinese strategy adopted an import-substitution approach, where the "infant industry" would produce for the Chinese market first with some government protections and then gradually begin to export when it reached an international level of competitiveness. There was no domestic market in China at the time that these firms began production—they had their eyes on Germany, Spain, Italy, and the United States.

The central government in China did not place great emphasis on solar PV technology in its RD&D policies or funding during the ninth, tenth, or eleventh five-year plan, nor did it establish substantial market-formation policy support during this period, unlike in the case of wind energy, where the government had experimented with many policies like the Ride the Wind policy, the major wind concessions, and so forth. The earliest R&D on solar PV technology in China reportedly commenced in 1958, and this initial PV research was centered mainly on the role solar PV could play in space applications. The same was true in the United States. Production equipment for solar cell manufacturing was first imported into China around 1985 (Yang et al. 2003).

Although the central government was not focused on solar PV industrial development until the mid- to late 2000s, the provincial and local governments were quick to get behind the new solar firms in their jurisdictions in hopes of creating local jobs and new industries as well as boosting local GDP. The city of Huai'an in Jiangsu Province, for instance, provided a 50 percent refund in the real interest of loans used to purchase equipment for a factory, a refund equal to 0.05 RMB/kilowatt-hour (kWh) in electricity consumption for the first year, a refund of the land transfer fee, and a partial refund of corporate income tax for the first eight years (Grau, Huo, and Neuhoff 2011).

Later, after the Chinese manufacturers had already successfully brought the costs of solar PV modules down, the central government began to provide domestic market-formation support through new policies. There are three main market-formation policies for solar PV today in China: Golden Sun, Solar Roofs, and Large-Scale On-Grid PV. The Golden Sun program, initiated in 2009, is designed to support the demonstration of key PV technologies, mainly in commercial buildings, remote rural residential buildings, and (originally) large-scale on-grid PV. Through a public bidding process, investors are subsidized for 50 to 70 percent of the cost. The Solar Roofs program supplies a subsidy for installing rooftop and building-integrated PV. The feed-in tariff for large-scale on-grid PV is decided through a bidding process. These policies are discussed in more detail in chapter 4.

Access to foreign technology was not a barrier for any of the emerging Chinese firms, as predicted by John Barton (2007) based on the fact that this industry is not oligopolistic. The entire Chinese solar industry was born through the licenses of foreign technology along with the importation of equipment and machinery, mostly from Europe.[24] As of 2003, there were seven main solar cell manufacturers in China with a total capacity of 7.15 MW peak, and all but one imported most, if not all, of the equipment to produce the technology (Yang et al. 2003). This initial technology transfer process occurred with remarkably few barriers and with incredible speed.[25] One of the leading solar PV manufacturers imported 80 percent of the equipment for the solar cell production, most of it from Germany. For polysilicon production, most of the equipment came from foreign providers, mainly in the United States. One manufacturer estimated that Chinese equipment providers were one-third to one-half the price, but their quality was so bad that it wasn't worth the cost savings.[26] During the development of China's automobile industry, the Chinese government stipulated local content requirements, which forced foreign manufacturers to either form joint ventures with Chinese parts manufacturers or train them to be able to meet performance requirements. Foreigners complained bitterly about the local content requirements, but the policies were extremely effective at improving the capabilities of the Chinese parts manufacturers (Gallagher 2006a). In the case of the Chinese solar industry, no local content requirements were permitted due to China's membership in the World Trade Organization (WTO).

Most of the top Chinese solar PV firms cultivated relationships with foreign research institutes or firms to conduct R&D in addition to acquiring firms outright. Suntech worked closely with the University of New South Wales, and New South Wales professor Stuart Wenham became the CTO of Suntech. Yingli formed contracts with the Energy Research Centre of the Netherlands to improve the efficiency of mono-silicon cells, and the funding for this research cooperation was split, with Yingli paying two-thirds and MOST the remainder. JA Solar signed an agreement with Innovalight, a firm based in Sunnyvale, California, to codevelop solar cells with conversion efficiencies exceeding 20 percent and also for Innovalight to provide silicon nanoparticle ink. Trina Solar and DuPont agreed in 2012 to begin collaborating on R&D efforts to advance the efficiency and lifetime of solar cells and modules.

The source of the Chinese firms' comparative advantage is a continuing topic of controversy. Some US and German firms (most vociferously the German firm SolarWorld, which has a manufacturing facility in the United States) have argued that the Chinese government has unfairly subsidized its firms. Led by SolarWorld, the US-based firms filed a complaint with the US Commerce Department in 2011 to try to trigger a process that would lead to the imposition of tariffs on Chinese imports. A different set of US firms in the Coalition for Affordable Solar Energy emerged to counter the first consortium out of concern about potential job losses in the US solar raw materials supply and installation industries if Chinese imports became more expensive. The US Commerce Department concluded that this support was equivalent to 2.9 to 4.7 percent of the value of the products. In May 2012, the Obama administration imposed preliminary tariffs on solar panels imported from China, and then finalized import tariffs in November 2012 that ranged from 18.32 to 249.96 percent. The European Union launched an inquiry as well, again spurred by SolarWorld.

There is no doubt that the Chinese firms received various kinds of support from the local governments, especially including low-interest loans along with discounted land and electricity prices, and this was confirmed by interviews.[27] On the other hand, US and German firms also received government support for their manufacturing plants.[28] Indeed, the US government has provided investment tax breaks, production tax breaks, and loan guarantees to many clean energy firms at the federal level, and it is not uncommon for further incentives to be supplied at the local level as well. In the SolarWorld case, SolarWorld vice president Bob Beisner said the company's move to the US state of Oregon was made possible by the state's business energy tax credit for renewable energy producers. The tax credit is awarded for 35 to 50 percent of the total cost, depending on the project, for the first twenty-two million dollars in project costs. Beisner was quoted in the local newspaper, stating, "It really put us over the edge. It was a key factor" (Gaston 2009). Similarly, thin film producer First Solar said that its decision to place its Agua Caliente plant in Mesa, Arizona, was the result of "supportive state and federal policies," according to then president Bruce Sohn (First Solar 2011). Clearly, governments in every country at the national, state, and local level are offering "carrots" to firms to locate and prosper there. It's hard to assess quantitatively whether or not the Chinese are doing it to a greater extent than other countries, or if they are simply doing it more effectively. But the focus on subsidization may be distracting from more fundamental sources of Chinese competitiveness.

In interviews, a much more nuanced picture emerged about the sources of Chinese competitiveness. Labor costs are not a significant factor in the Chinese solar PV manufacturing industry since the manufacturing processes are highly automated, even in China. Factory visits in 2010 confirmed the scarcity of workers on the assembly lines due to widespread automation. Several factors appear to be important, starting with the flexibility of Chinese manufacturers. According to one foreign manufacturer, the speed at which the Chinese can react to orders and other changes in the marketplace is unrivaled. Part of this flexibility derives from the cluster effect of parts and components suppliers, most of which are located in eastern China. Many ancillary equipment manufacturers are close by, and these firms are also able to turn on a dime and alter production as needed. Due to less protective labor laws, the Chinese can also ramp production up and down in immediate response to the market, sending home workers if need be and then rehiring them when the market picks back up.[29]

The Chinese's culture of frugality was also cited as strongly affecting their ability to reduce costs in the manufacturing process. The Chinese manufacturer's attitude is always how to reduce costs.[30] In a speech at MIT in 2010, Suntech CEO Shi Zhengrong said that when he wrote the business plan for Suntech, he had no idea how much everything would cost, so he decided to estimate a 30 percent discount on everything. He was confident it could be done and stated that his philosophy was "thoroughness." The early shortages of silicon also inspired Chinese firms to use it more efficiently.[31] One firm noted that it focused heavily on how to make the wafer thinner so as to use less silicon.[32] During a tour of one manufacturing plant, I paused to watch a camera flash over each finished wafer to determine its efficiency, and the cell efficiency of most cells was about 16.5 percent, with approximately 10 percent of the wafers higher than 17 percent efficiency. I murmured compliments, which were immediately and forcefully rebuffed as my host declared that the efficiency still wasn't good enough and the goal was to achieve at least 20 percent efficiency within a few years. As a representative of a foreign firm commented in an interview, "The Chinese have done a good job with manufacturing."[33]

A different explanation of the Chinese's ability to reduce costs in the manufacturing process more recently is the ruthless and fierce competition in the Chinese market. As is the case in many other industries in China, success breeds copycatting, and many new firms tried to enter the marketplace after the initial successes of the major firms like Trina, JA Solar, Suntech, and Yingli. Local governments support these new firms in hopes of boosting job creation and local GDP. This phenomenon breeds "repetitive" production. Estimates of the number of solar PV module manufacturers in China vary, but there are at least one hundred manufacturers even though the top ten biggest firms dominate the market. Still, the 2011–2012 downturn forced consolidation; by one account, 50 percent of Chinese manufacturers had suspended production as of December 2011 (Chang 2011).

Another final explanation for Chinese competitiveness is the vertical integration of many Chinese firms. Two interviewees at different companies raised this factor,[34] In addition, 2011 data provide evidence that the Chinese firms have indeed overwhelmingly opted to vertically integrate in order to control the full production process and remove margin stacking. Three-quarters (eight out of eleven) of the Chinese firms in the top twenty producers globally are fully vertically integrated (see table 3.2).

The availability of finance for firms operating in China was not identified as a constraint by any interviewee, Chinese or foreign. Several remarked that it was easy to attract financing and the transaction costs of doing business in China were fewer than elsewhere. Of course, these firms quickly became hot commodities. As one interviewee said, "Success breeds success."[35] Once a little finance is acquired, more and more can be accumulated. A different Chinese firm commented that it could get capital in many ways, from the banks, the stock market, and also the owners themselves.[36] At least one foreign government, Germany, has also provided low-interest loans to the Chinese solar industry. The

Table 3.2
Top twenty solar PV module producers

Firm	Nationality	Type of integration	2011 production (MW)
Suntech Power	China	Partial	2,010
First Solar	United States	Thin film	1,981
Yingli Green Energy	China	Full	1,684
Trina Solar	China	Full	1,510
Canadian Solar	Canada	Partial	1,386
Sharp	Japan	Partial	986
Sunpower	United States	Full	964
Tianwei	China	Full	938
Hanwha-SolarOne	Korea	Full/thin film	928
LDK Solar	China	Full	880
Haeron Solar	China	Partial	855
JA Solar	China	Partial	820
Jinko Solar	China	Full	792
Kyocera	Japan	Full	660
REC	United States	Full	644
SolarWorld	Germany	Full	605
Jiawei SolarChina	China	Partial	598
Astronergy	China	Partial	550
Changzhou Eging PV	China	Full	525
Aide Solar	China	Partial	497

Source: GTM Research, http://wwww.greentechmedia.com (accessed April 27, 2012); Wesoff 2012.

Kreditanstalt für Wiederaufbau Development Bank gave a loan to the Export-Import Bank of China, which added funds that in total were then lent to Yingli, Sunergy, JA Solar, and LDK Solar. The latter then reinvested in a German solar producer, Sunways (Neubacher 2012).

As for intellectual property, no court cases exist related to solar PV infringement in China to date, and interviews revealed no rumors either. One foreign company official went so far as to state, "I definitely think there is a myth around protecting IP. . . . [I]n solar technology . . . much of the knowledge is generalized, and . . . you can pretty quickly figure out how to protect yourself."[37]

As for the ability of the Chinese to acquire technologies as well as export them, the barriers have not been insurmountable. Polysilicon was initially imported (and still is for some firms). One Chinese firm said it had trouble acquiring licenses for polysilicon production because US and Japanese firms had refused to license it, so they either had to import polysilicon or rely on inferior Chinese technology for the time being.[38] On the other hand, Chinese firms have aggressively made strategic acquisitions and investments into foreign firms and capabilities as needed. To take the example of Suntech, it acquired MSK, a leading supplier of building-integrated PV systems, to gain access to building-integrated PV technology as well as wider entry into the Japanese market. It acquired EI Solutions, a commercial solar systems integration company in the United States, as well as two investment funds to develop, finance, and own projects in the United States. Suntech also invested in upstream suppliers to secure high-quality and low-cost polysilicon and silicon wafers, including acquisitions of minority stakes in Asia Silicon, Glory Silicon, Hoku Scientific, Nitol Solar, and Xi'an Longji Silicon. It acquired strategic assets, too, in order to enhance its manufacturing and design capabilities, including German KSL-Kuttler Automation Systems and CSG Solar, another German firm involved in crystalline silicone technology. All the Chinese firms proudly tout the number of patents they already hold.

Coal Gasification for Electricity Production

The story of the development of coal gasification technology in China is an exemplary case of technological "catch-up." In coal gasification technology, the Chinese originally acquired the technology through direct product purchases, some of which did not work. Due to poor after-sales service on the part of the providers, the Chinese were forced to try to learn why the gasifiers were not working in China, and through much trial and error, they figured out how the technology worked and improved on it. It is now widely acknowledged that the Chinese have caught up to Western firms in this domain. Firms in the United States have begun to license the Chinese gasification technology. Although the Chinese perceive the learning process to have been long, arduous, and painful, it was effective. The initial barriers were high, but the Chinese managed to overcome them in the end, emerging as global leaders in this technology.

Also known as coal conversion, coal gasification is a process where coal is fed into a gasifier, subjected to high pressure and temperature, and converted into different chemicals and fuels, including synthetic natural gas. China is the largest user of coal gasification technology in the world. The advantages of coal gasification for a country like China are many. Gasification technology is relatively energy efficient as a coal-fired power generation technology, it can produce different types of products (including chemicals), and it is relatively easy to separate and capture CO_2 after the coal is gasified. If the CO_2 is captured, it can be reused for other industrial applications, or sequestered in underground geologic formations or depleted oil and gas reservoirs.

Although there are no commercially operating plants that capture and sequester CO_2 in China now, the government is supporting a number of smaller-scale demonstration plants. Huaneng is also building a large-scale IGCC plant called GreenGen near Tianjin, which will eventually separate and capture CO_2 in a later phase. Through MOST and CAS, the Chinese government has put great effort into conducting RD&D on advanced and cleaner coal technologies since the 1980s.

Coal gasification technology was originally introduced into China through the import of foreign gasifiers, largely for fertilizer production. The Chinese purchased Lurgi pressurized fixed bed, U-Gas fluidized bed, and Texaco and Shell entrained-bed gasifiers (GE now owns the Texaco technology). The Texaco-GE gasifier utilizes a coal-water slurry feed, and the Shell gasifier utilizes a dry powder feed. Siemens later developed a coal gasifier that is similar to the Texaco-GE gasifier. When they were first introduced, the U-Gas and Shell gasifiers did not work well. The Texaco technology was not well suited to Chinese coal characteristics, so the Chinese had to import coal. Chinese scientists spent a long time confirming that the Chinese coal types didn't work well with Texaco technology.[39] The U-Gas technology was introduced to the Shanghai Coking Plant, and none of the eight gasifiers worked well. U-Gas technology was doomed after that because no one else wanted to buy expensive equipment that wouldn't work. Shell had a similar problem, and lost significant market share later because of it.[40]

The Xian Thermal Power Research Institute (TPRI) and East China University of Science and Technology (Huadong) both subsequently developed dry pressurized gasification technology. An important technical modification for both was to use multiple nozzles, and the Huadong gasifier has achieved greater efficiencies than the Texaco one (no data on efficiency were available for the TPRI gasifier).[41] These are the two leading gasifiers in China today, but others have been developed at Tsinghua University and elsewhere.

When Siemens came to China it established a joint venture to produce coal gasifiers with Ningmei, which is affiliated with Shenhua Group, and they worked together on initial projects, which helped to ensure that the technology was adapted and suitable for the Chinese market, according to one expert. For years GE opted not to form a joint venture for gasification, preferring instead to sell final products or license the technology, but the license fees are expensive, and GE began gradually losing market share in China. The same Chinese expert observed, "The Texaco technology is so old now, why are they working so hard to protect it?"[42] In 2012, GE announced it was forming a fifty-fifty joint venture with Shenhua Group to sell industrial gasification technology licenses in China, conduct R&D, and market IGCC solutions in China.

Several Chinese scientists confirmed that the license fees for the foreign gasifiers were and remain high.[43] The fees were not so high that the Chinese couldn't afford them, because after all, dozens of gasifiers were imported. On top of the license fees, however, there were technical service fees, training fees, assessment fees, and all these could be doubled depending on site conditions. The Chinese initially failed to negotiate guarantees for the technologies, which the Chinese now believe was a huge mistake. On the other hand, Chinese experts say that the high license fees and bad after-sales service motivated the Chinese to develop their own technology, based on their detailed knowledge about Chinese coal characteristics.[44]

As will become clear in great detail in chapter 5, no formal court cases of IP infringement have been filed related to coal gasification technology in China. Yet some people believe that some infringement did occur when Texaco first formed a partnership with the Northwest Research Institute in China because they did some joint demonstrations, and the institute learned a lot from this experience. Based on this experience, the Northwest Research Institute then made further modifications on its own.[45]

In terms of technology policy, MOST introduced "clean coal" technology into the tenth five-year plan (2001–2005) for the first time, aiming to strengthen indigenous innovation in advanced coal technology. The strategic goals of the tenth five-year plan were to produce coal-based liquid fuels, alleviate oil shortages, increase coal utilization efficiency, and reduce pollution. Coal gasification was a key target technology to help achieve all these goals. Prior to the tenth five-year plan, though, the Chinese government had initiated research on coal gasification as early as the 1970s, when it decided to build an experimental ten MW IGCC power plant. The government continued to support R&D on IGCC during the subsequent decades. In 1999, the government approved a large-scale IGCC demonstration project in Yantai, Shandong Province, with an intended capacity of three to four hundred MW. But when the project went out to bid, the costs were much higher than anticipated due to the need to import nearly all the technology and equipment, and the project was halted. No other commercial-scale IGCC plants were approved until the GreenGen project, but many coal gasification plants for liquid fuels and/or chemicals have proceeded during 1990s and 2000s (Zhao and Gallagher 2007). The GreenGen project is using a Chinese gasifier, developed by the TPRI. Shenhua Group is one of China's most aggressive developers of gasification projects, with a direct coal liquefaction plant in Ordos and many other places.

While there is a natural market for coal gasification technology in China due to the high demand for chemical products and liquid fuels in China, there is no natural market for IGCC technology for electricity due to its relatively high costs compared even with the most efficient pulverized coal technology, ultra-supercritical coal. A 2010 study put the first-of-a-kind demonstration costs of IGCC in China at 7,751 RMB/kW (kilowatt), with a reference electricity price of 504 RMB/MW-hour. By comparison, ultra-supercritical coal with flue-gas desulfurization (FGD) had a unit cost of 3,924 RMB/kW, and regular supercritical PC with FGD was 3,919 RMB/kW, with a reference electricity price of 321 RMB/MW-hour, largely confirming an earlier study (Zhao et al. 2008; Chen and Xu 2010). Of course, these first-of-a-kind costs will almost certainly come down as additional units are built. Also, these costs do not reflect any of the environmental benefits that accrue from use of IGCC technology. IGCC technology emits little sulfur dioxide without the need to use FGD or CFB technology, and it emits the least nitrogen oxide and particulate matter of any of the coal-fired power-generating technologies. As previously mentioned, it is relatively easy to capture CO_2 from IGCC plants, although there is an efficiency penalty, which further increases the cost.

Due to all the environmental and health advantages of IGCC technology over other coal technologies, and the strong likelihood of future cost reductions, there is ample justification for the government to provide preferential support for IGCC plants in China. Indeed, the research agencies and ministries including MOST and CAS have provided substantial R&D support for decades, but a lack of coordination between them and the NDRC becomes apparent when one notices that only the GreenGen plant has been approved for construction. At least a dozen IGCC proposals have been submitted for approval by the NDRC, but according to many sources, the NDRC remains concerned about the costs of IGCC. One senior official who headed the Energy Bureau for many years, Zhang Guobao, was personally unconvinced about the merits of IGCC technology, and blocked the approval of these projects until his retirement. Thus, the market-formation policies of the Chinese government for coal gasification for electricity have been virtually nonexistent. This problem has led companies and research institutes to conclude that they must do IGCC as a "coproduct" of liquid fuel or chemical production, since they have more value in the marketplace.

The GreenGen project is significant because of the technology transfer and international cooperation components. The idea of what is now GreenGen grew out of China's knowledge of the FutureGen project in the United States. FutureGen was an IGCC demonstration project planned by the US Department of Energy and an industrial consortium, the FutureGen Alliance. International partners were welcomed, and Huaneng joined the consortium in 2004. The idea was to build a state-of-the-art commercial-scale IGCC plant that would be coupled with CCS. Huaneng and China's MOST were inspired by the idea, and hatched the notion of developing a version of FutureGen in China, what is now GreenGen. The US FutureGen project was halted by the George W. Bush administration, briefly resurrected by the Obama administration, and is essentially halted today. Ironically, China through its GreenGen project will likely beat the United States in demonstrating a coupled IGCC-CCS plant.

GreenGen has nine shareholders, with Huaneng holding 52 percent of the shares, and all the rest each holding 6 percent. The other shareholders include major energy companies Datang, Guodian, Huadian, the China Power Investment Corporation, China National Coal Group, State Development and Investment Corporation, Shenhua Group, and the US firm Peabody Coal. Separately, the US firm Duke Energy has formed a collaborative relationship with Huaneng. The Chinese government has provided support for the project as well, valued by one source at 350 million RMB (approximately US$46 million) (MIT 2012). GreenGen will be a 4,000 MW plant using the TPRI gasifier, which is being significantly upscaled for this project to 2,000 t/d from its previous demonstration project, which was only 36 t/d. It is using a Siemens E-class gas turbine and an air-separation unit from Kaifeng. In the first phase, the main purpose is to demonstrate the TPRI gasifier at this scale. GreenGen estimates that the TPRI gasifier is 20 percent cheaper than the Shell one, mainly due to the avoidance of license fees.[46]

Barriers to and Incentives for the Cross-Border Transfer of Cleaner Energy Technologies

Taking these four case studies as a whole, it is possible to analyze the extent to which different kinds of barriers and incentives have inhibited or facilitated the cross-border flow of cleaner energy technologies to and from China. Tables 3.3 and 3.4 below provide a synoptic picture of the different barriers and incentives from both foreign and Chinese perspectives based on the four empirical case studies. Where Chinese identify a barrier or incentive, the color is black. Where foreigners identify a barrier or incentive, the color is gray. Where Chinese and foreigners agree, there are stripes. A question mark denotes a lack of data, whereas a lack of entry means that the barrier or incentive does not apply in that case.

To begin with policy, in table 3.3, one can observe *widespread agreement* between Chinese and foreign experts on the incentives that have encouraged the cross-border transfer of these four cleaner energy technologies to and from China. Most decisively, all types of policy including clear targets over time, lack of significant barriers to trade and foreign direct investment (FDI), strong innovation policy, stable market-formation policy, strong export promotion policy, and alignment of all these types of policy were identified as important incentives for technology transfer. None of these policy incentives applied in the gas turbine case because there has been little concerted policy to encourage the development and deployment of gas turbines to date in China, and consequently, few have been deployed. Fewer of the policy incentives applied to coal gasification because the development of this technology in China has largely been "pushed" by MOST, with little "pull" in the form of market-formation policy from the NDRC.

In striking contrast, there is *little agreement* between foreigners and the Chinese on the policy barriers, as depicted in table 3.4. Foreign firms identify a lack of access to the Chinese market as a key barrier in both advanced-batteries and solar PV technologies. Chinese experts believe that Chinese policy has been inadequate in all cases except for solar PV industrial policy and innovation policy for coal gasification. Chinese solar PV firms have benefited from provincial and local industrial policies and programs, but there is no formal central government industrial policy for the solar industry. Both Chinese and foreigners believe that Chinese market-formation policy is rapidly improving in the support of batteries and solar PV. The Chinese identify barriers across the board for

Table 3.3
Incentives for the international transfer of cleaner energy technologies to and from China

		Gas turbines	Advanced batteries for vehicles	Solar PV	Coal gasification
Policy factors	Clear targets historically and long-term policy		*	*	
	Lack of barriers to trade and FDI				
	Strong innovation policy				
	Strategic industrial policy		*	*	
	Stable market-formation policy		*	*	
	Strong export promotion policy				
	Alignment of all types of policy				
Cost and finance factors	Good access to finance				?
	Natural market exists				
	Ability to buy technology if needed				
	Costs of foreign or Chinese technology reasonable				
IP factors	Strong or improving patent regime domestically				
	Confidence in domestic courts	Some	Some	Some	Some
	Willingness of foreign firms to license or cooperate in joint development	Some	Some		
	Strong domestic technological capabilities				
	Knowledge of technology needed/absorptive capacity				
Business practice factors	Experience in foreign markets				
	Flexibility and nimbleness of firms	Some			
	Colocation with supply chain				
	Global perspective on markets				
	Good IP management				
	Tolerance for risk taking	Some	Some		Some
	Good after-sales service	?			

Notes: Black is Chinese point of view. Dark gray is foreign point of view. Light gray denotes agreement between foreign and Chinese perspectives. An asterisk (*) means taking into account new policies in the twelfth five-year plan, but not necessarily historically. A question mark (?) denotes a lack of data, whereas lack of an entry means that the barrier does not clearly apply in this case.
Source: Author analysis, based on case study research.

Table 3.4
Barriers to the international transfer of cleaner energy technologies to and from China

		Gas turbines	Advanced batteries for vehicles	Solar PV	Coal gasification
Policy factors	Export controls				
	Import tariffs				
	Restriction of access to domestic market				
	Weak innovation policy				
	Weak industrial policy				
	Weak market-formation policy				
	Weak export promotion policy				
Cost and finance factors	Access to finance/ ability to invest				
	Lack of "natural" market				
	High cost of foreign technology				
IP factors	Export prohibitions in license agreements			?	
	Defensive, anticompetitive patenting			?	
	Fear of IP infringement				
	Refusal by foreign firms to license				
Business practice factors	Lack of experience in foreign markets				
	Weak IP management				
	High risk aversion				
	Poor after-sales service				

Notes: Black is Chinese point of view. Dark gray is foreign point of view. Light gray denotes agreement between foreign and Chinese perspectives. A question mark (?) denotes a lack of data, whereas lack of an entry means that the barrier does not clearly apply in this case.
Source: Author analysis, based on case study research.

gas turbines, and the only points of agreement with foreigners about this technology are the lack of a "natural" market in China and risk aversion due to extreme market uncertainties. The foreign perception of Chinese policy is largely positive because foreigners believe that the Chinese government is committed to clean energy, and likely will be able to implement and sustain stable market-formation policies in the future. Foreigners generally believe that the Chinese government takes a longer-term view.

Cost and financing factors present different barriers for Chinese versus foreign firms. Chinese firms have good access to capital in the battery and solar PV industries, although foreign firms cited this factor as a major barrier for them at home. There seem to be no barriers for firms to license technology or acquire firms in either direction, except for the cases of gas turbines, HEVs, and polysilicon production. Chinese firms routinely license technology from foreign firms and have even purchased foreign firms outright in order to acquire technological knowledge. Foreign firms have licensed technology from the Chinese as well, most notably in coal gasification. And foreign firms have formed joint ventures and acquired Chinese firms. The Chinese and foreigners do not see eye to eye on the costs of foreign technology, which the foreigners view as legitimate given their R&D investments and business practices, and which the Chinese generally see as too high (except in the case of solar PV). All agree that the Chinese gasifiers are now highly competitive in the marketplace.

Foreign firms routinely identified lack of capital as a barrier for expansion activities, particularly outside China. The high cost of foreign technology was a key barrier for the Chinese in the cases of gas turbines, advanced batteries, and coal gasification. But these high costs motivated the Chinese to develop their own indigenous technology in coal gasification and advanced batteries, and in the case of coal gasification, the Chinese have already caught up and become globally competitive. Everyone agrees that the lack of a "natural" market for all four technologies is a problem that must be overcome through government intervention. In other words, because the marketplace does not value the benefits of the technologies in terms of reduced greenhouse gas emissions, conventional pollution, public health, or energy security, the cleaner technologies have a harder time competing against incumbents.

In the area of intellectual property, there was clear agreement among Chinese and foreigners alike that there is a strong or improving patent regime in China. Confidence in Chinese courts is cautiously growing, with some firms believing that the courts are pretty fair, a source of potential remedy, and worth utilizing, and others still suspicious. Chinese firms were more suspicious of the courts in situations where Chinese would be suing other Chinese. At least some foreign firms are willing to license technology to China in most cases, which means that even in situations where one firm refuses to license, there usually is a second option, and therefore Chinese firms unambiguously have access to cleaner foreign technology, with the possible exceptions of certain hybrid vehicles and the most advanced gas turbine technologies. There is evidence that some foreign firms do indeed refuse to sell or license certain cleaner energy technologies to their Chinese counterparts. The Chinese have extremely good knowledge of what technologies they need to acquire and reasonably good absorptive capacities. They have excellent project execution and manufacturing capabilities in all cases, so they can get technologies into production as needed. Design capabilities remain weak in all the technologies except for coal gasification.

The Chinese observed more barriers in IP than foreigners have found because, at times, they experienced export prohibitions in license agreements, defensive patenting, and plain refusals to license technology or final products in gas turbines as well as advanced-battery technologies. Solar PV firms have also experienced license refusals from foreign firms for polysilicon production, though these refusals do not seem to have substantially harmed the Chinese solar PV industry. For every technology except coal gasification, foreign firms still fear IP infringement in China.

Neither Chinese nor foreign firms identify many barriers in the area of business practice. Many different practices create incentives for technology transfer. Foreign firms active in the Chinese market all noted their long experience working there and pointed out that less experienced firms might make significant mistakes. The flexibility and nimbleness of Chinese firms in manufacturing batteries, solar PV, and coal gasification were cited as significant comparative advantages for the Chinese by both Chinese and foreign observers. Where foreign firms were active in China, the cost advantages and flexibility of manufacturing there were cited as being important factors in their decisions to bring technology there. A significant comparative advantage for the Chinese firms is their colocation with the supply chain in China, although many Chinese firms source inputs from foreign firms in the battery and solar PV industries. A global perspective on markets clearly facilitates international technology transfer in both directions. This perspective is most pronounced in the solar PV sector, where the industry grew primarily through a reliance on exports, but this perspective is also observable in advanced batteries and coal gasification. In all cases, foreign firms had a strong global perspective about being active in China, except for some instances in coal gasification where risk aversion was strong or grew among some firms. In these four case studies, the foreign firms all utilized good IP management practices to protect their IP, and while annoying and frustrating problems had arisen, none of the firms had experienced life-threatening IP infringement. As will be explored in chapter 5, there is one case in another clean energy sector—wind—that may prove to be significant, which is the dispute between American Superconductor and Sinovel. Both firms have filed court cases against the other, and the case is unresolved as of this writing. Some foreign firms have a high tolerance for risk taking, and they are being largely rewarded for that in the Chinese marketplace. Chinese solar PV firms that imported vast quantities of manufacturing equipment from abroad had no complaints about the after-sales service of their suppliers. This factor was a big problem with early coal gasification licenses, but now the new Chinese gasification suppliers are developing a reputation for being active problem solvers.

Conclusion

The four tales told in this chapter provide new empirical evidence about the barriers and incentives for the cross-border transfer of cleaner energy technologies to and from China. Overall, no insurmountable barriers to the acquisition or export of clean energy technologies were identified. Although some foreign firms refused to license or sell certain technologies to Chinese firms, a second option could almost always be found. Domestic policy was found to be absolutely crucial to the creation of structured incentives to facilitate international technology transfer, in both directions. Chinese government policies created incentives for Chinese firms to acquire cleaner technologies from abroad and for foreign firms to bring their technologies to China. Domestic policies in Germany, Spain, and the United States especially created incentives for Chinese firms to acquire foreign technology, manufacture, and export to these markets. Other major incentives for the cross-border transfer of technology were good access to capital and smart business practices.

National policy creates the most important incentive for the cross-border transfer of cleaner energy technologies. There are at least three different types of policy that have special roles to play in deploying cleaner energy technologies: innovation, industrial, and market-formation policies. These policies are discussed in detail in chapter 4. Yet policies that restrict access to the domestic market unequivocally hinder the international technology transfer of cleaner energy technologies.

The prices of cleaner foreign technologies were often deemed to be too high by most Chinese, but the Chinese usually paid these prices anyway while simultaneously beginning to innovate to avoid having to pay such high prices in the future, indicating that foreign firms might need to reconsider their pricing strategies if they are only catalyzing Chinese firms to develop their own technologies. In the battery, solar PV, and coal gasification cases, the Chinese have contributed to global reductions in the final costs of cleaner energy technologies, as explored more thoroughly in chapter 6. The new Chinese competitor firms have thus shifted the global supply curve to the right, which in turn stimulates greater demand for cleaner energy technologies due to their lower prices. A major reason why the Chinese have been able to buy the technologies they need and manufacture efficiently is their seemingly unrivaled access to capital. Chinese firms are not constrained by difficulties financing expansionary activities, unlike most foreign firms, even in advanced industrialized countries like the United States and Germany. Access to finance is therefore a crucial facilitator of international technology transfer.

The cases shed new light on the two-sided coin of intellectual property rights in cleaner energy technologies. In these four cases, no evidence of serious infringement of foreign technologies in China was found, although there were a couple of instances of rogue employees who tried to personally gain from knowledge they acquired working for firms. There was surprising praise for growing IP protection in China and the professionalism of China's patent office. Confidence in the courts seems to be growing among foreign firms, though they remain fearful about being able to obtain injunctions and receiving a fair hearing. As described above, Chinese firms have demonstrated resolve and creativity in finding foreign firms and consultants to help them acquire as well as implement needed technologies, and were almost always able to get the technologies that they wanted from one firm or another. These issues are examined in much more depth in chapter 5.

Finally, a global perspective on markets proved central to the ability and willingness of firms to explore foreign markets, leverage market-formation policies around the world, and avoid excessive risk aversion. This is true of both Chinese and foreign firms. Gradual acquisition of experience was imperative for all firms, either through study or work opportunities in other countries, or initially through smaller business ventures abroad that represented little risk to the firm. The following chapters take up each of these major issues in turn.

4

The Essential Role of Policy

Government policy is extremely important to drive long-term sustainable development.
—Ed Lowe, GE

Without government regulation, you won't have a market for clean energy.
—Hans-Peter Böhm, Siemens

The policy environment is important—principally the stability and the predictability.
—Richard Wilder, Microsoft

Government policy strongly affects the deployment of cleaner and more efficient energy technologies around the world. This chapter explores how national and subnational policies have had the greatest effect to date, although international policies have certainly had a modest effect, primarily through catalyzing national policymaking. Policies such as feed-in tariffs, performance standards, renewable portfolio standards, energy and carbon taxes, and cap-and-trade instruments have been enacted and implemented in almost every country. One hundred and nine countries have enacted some form of policy to support renewable power, and 118 countries have set targets for renewable energy. Ninety-two states, provinces, or countries have feed-in policies, and 71 have either a renewable portfolio standard or quota policy (Ren21 2012). Aside from in Europe, there is little coordination among countries on these policies and few attempts at formal harmonization. The global policy landscape is thus a mosaic of many different types of policies that cumulatively affect the global marketplace for clean energy. The heterogeneity in the policies has created a somewhat haphazard global market for producers, but it has also allowed for considerable policy experimentation as well.

The countries supporting renewable energy are not just "green" European countries but include middle-income and developing countries, too. China and Turkey have the number one and two largest solar hot water capacity in the world, respectively. China, Vietnam, Brazil, and India hold the top four spots in hydropower capacity. China, the United States, and India are the top three in wind power capacity. And the United States, Brazil, and China are the top three ethanol producers. It is notable that China holds the top spot in terms of new renewable capacity investment, hydropower capacity, wind capacity, and solar hot water capacity as of 2012. The United States ranks second in new capacity and wind power capacity, and first in biodiesel and ethanol production. Surprisingly, given its size, Italy has the largest solar PV capacity. Modern renewables now account for 8.2 percent of the global energy consumption, and traditional biomass accounts for another 8.5 percent (Ren21 2012).

Motivations for the creation of these policies are unique to each country, but four factors seem to influence a government's decision to adopt favorable policies for renewable energy: economic motives, a high endowment of renewable resources and/or a low endowment of nonrenewable energy sources, the political system, and cultural factors and attitudes (Gallagher 2013). Economic motives include seeing the potential for new jobs in an emerging industry that is growing quickly and is likely to be dominant in the coming decades. The International Energy Agency (IEA) estimated that the needed cumulative investment in energy supply infrastructure is US$38 trillion (2010 dollars) for the 2011–2035 period. Renewable energy accounts for 60 percent of the projected new investment in the power sector. The share of world trade accounted for by energy is also growing rapidly, accounting for 13 percent in 2005 and projected to rise to 16 percent by 2016 (IEA 2009). Countries with high renewable endowments, such as solar insolation or wind power, are eager to exploit them, whether they are Laos or Canada. Conversely, countries that are energy resource scarce, like Japan and Denmark, are embracing energy efficiency and renewable energy to improve their energy security. Some political systems have allowed green constituencies to influence policies more than others. And finally, cultural factors, while impossible to quantify, seem to affect attitudes differently around the world.

Through the case study research and additional interviews with government officials and experts, government policies in the United States, China, and Germany were examined in some detail. In addition, major market-formation policies at the national and subnational level around the world were documented through a variety of sources for the period 1970 to early 2012. These policies are presented in a timeline in appendix D. The case studies and interviews were used to identify which types of policies affect the global diffusion of energy technologies, and how.

Policy Debates and Dilemmas

There are three primary debates that revolve around the rationale for policies in support of cleaner and more efficient energy technologies: economic, environmental, and security. We can caricature them as politicians often do, as "how to achieve a green economy," "how to solve climate change," and "how to achieve energy security."[1] There are serious questions about how far reaching the goals should be, how to cost-effectively achieve the goals, how to make sure the costs and benefits are fairly distributed, and which policies would actually achieve the desired outcomes. In this section, each of these debates is examined with a global perspective.

The Green Economy
Most countries have stated an interest in developing a green economy, or following a sustainable development pathway. Much has been written about how the next "revolution" following the current information revolution may be a clean energy one, and how such a transition could not only be good for the environment and energy security but also might bolster a nation's economic competitiveness and help with job creation. According to President Obama (2011), "As we recover from this recession, the transition to clean energy has the potential to grow our economy and create millions of jobs—but only if we accelerate that transition. Only if we seize the moment." Likewise, former premier Wen Jiaobao (2009) remarked,

China ranked number one in the world in terms of installed hydropower capacity, nuclear power capacity under construction, the coverage of solar panel of water heater and cumulative installed photovoltaic power capacity, and fourth in the world for installed wind power capacity. These are major achievements in China's efforts to adjust economic structure and transform development patterns. They also contributed positively to the global endeavor to develop green economy and tackle climate change.

Perhaps only in the United States are the policy debates so fierce about whether and how to spur a clean energy economy. Many countries commonly develop industrial policies for certain sectors, and in fact, this was also done in the United States in the late nineteenth and early twentieth centuries (Chang 2002). But with the increased dominance of the "free market" ideology in the United States, resistance to market intervention has grown strong in the United States, ironically at the same time that protectionism has reared its head. The United States is also unusual in its inability to formulate a strategic national energy policy, whereas other countries have proven much more capable of setting national targets, and devising and implementing plans for achieving them.

Many countries support their firms in their quest to find and exploit foreign markets. It's worth examining how some of the leading countries have managed to achieve a high degree of export competitiveness in clean energy. Ninety-six countries currently have national renewable energy targets, but the United States is not one of them at the federal level, although more than thirty US states have implemented various forms of renewable portfolio standards (EIA 2012c; Sawin 2012). First, these countries set long-term targets for the domestic consumption of certain types of energy. In 1990, Denmark set a target that 10 percent of its electricity consumption should be supplied by wind by 2005, and recently revised its wind target to 50 percent of electricity consumption by 2030. In 1997, Germany set a target of 12 percent renewable electricity by 2010, but surpassed it early in 2007. In 2010, it revised its target for renewable electricity to 35 percent by 2020 and 80 percent by 2050. China has a national target of 15 percent renewable energy by 2020. Progress against these broad national goals can be measured, and policies can be designed to meet these goals. Second, these countries established stable policies to support the achievement of these aims. Denmark allowed individuals and residence-based cooperatives to produce wind, and starting in 1979, provided them with a 30 percent investment subsidy. People living within three kilometers of a turbine could invest in it through a cooperative, and this investment was tax deductible, as was the sale of surplus electricity to the grid (Buen 2006). By allowing anyone to become a renewable energy producer, popular support for wind energy increased enormously. The government required the utilities to purchase this renewable energy and set the rate at which it would be purchased to ensure that it would be profitable for producers. On the research side, the government supported R&D, and also created testing centers so that new inventions could be publicly tested and the learning shared with others (ibid.).

Some countries have created fiscal incentives for the production and use of efficient or renewable energy technologies. Japan is the most energy-efficient major advanced economy in the world. To encourage consumers to buy more efficient automobiles, it created a rating system and gave rebates to consumers who purchased the most efficient products (and imposed fees on buyers of inefficient products). Sure enough, consumers flocked to the efficient products, which motivated automotive producers to develop even more efficient cars. Japan also established strict, dynamic efficiency performance standards for many products including automobiles, electric toilet seats, computers, and air conditioners that manufacturers and importers are required to meet. In setting these standards through Japan's Top Runner program beginning in 1998, manufacturers were required to use the best-available efficiency technology, and the best-available technology is constantly updated to acknowledge new developments and products (Nordqvist 2006).

Germany's feed-in tariff system for renewable electricity was pathbreaking, with its earliest version beginning in 1991. Starting with its Renewable Energy Sources Act of 2000, the government stipulated that grid operators must accept and pay for the renewable energy fed into the grid, although the operators can pass costs on to the consumer. This law was most recently amended in 2012. The government sets "degression rates" so that the reductions in subsidies are predictable, although in 2010, the government unexpectedly reduced rates twice due to higher than expected installations and reduced installation costs. Of course, this feed-in tariff is not free; the electricity consumers of Germany pay for it, and feed-in tariffs have been criticized for being economically inefficient and oversubsidizing the renewables producers. But they are undeniably successful at providing a long-term signal to the market and creating a strong motivation for renewable energy firms to develop and deploy cleaner energy technologies (IPCC 2012; Butler and Neuhoff 2008).

Some countries directly support exports as well. Germany supports renewable energy demonstration projects abroad, showcasing German technology in various countries, which boosts exports. Some countries, such as Japan, tie their foreign aid to purchases of equipment from their own country. Others provide favorable loans and other support through Export-Import (Ex-Im) banks.

Explicit Climate Change Policy

All the policies described above and in more detail later in this chapter could be categorized as *indirect* climate change policies because they were not necessarily enacted with the goal of reducing greenhouse gas emissions even though they had that effect. Other policies are more explicitly aimed at reducing the threat of climate disruption. A number of countries have implemented carbon taxes. Some of these taxes, like India's, are small and mainly aimed at raising revenue for national investments into energy technology innovation. Others are revenue neutral, as in Switzerland, where revenues are redistributed proportionally to companies and households. Still others are designed to raise revenue in order to reduce corporate taxes or social security contributions (the United Kingdom, Denmark, the Netherlands, and Germany), or personal income taxes (Sweden and Finland) (Andersen 2012). All these types of carbon taxes create a direct incentive to reduce greenhouse gas emissions (Metcalf and Weisbach 2009). Another approach is to create a trading system for permits to emit greenhouse gases. In these systems, such as in the EU Emissions Trading System, the total emissions are capped, and the government issues permits to emit greenhouse gases to corporations either through an auction or free allowance system (or a hybrid) that add up to the total allowable emission cap. Revenues from auctioned allowances can be used for all the same purposes as carbon tax revenues: reduced personal or corporate income taxes, support for social security or health care programs, and investments in innovation.

Energy Security

At the human scale, energy security can be as simple as having access to energy resources for heating, cooking, refrigeration, and lighting. At the national scale, energy insecurity can be defined as the extent of vulnerability posed by the nation's energy system. Improving energy security goes far beyond achieving "independence" from oil imports because countries are vulnerable to attack on their energy infrastructure, including power plants, the grid, pipelines, refineries, and storage facilities (Lovins and Lovins 1982). Energy independence is an attractive term politically, but for most countries, it is difficult and costly to achieve, if even possible.

With respect to reducing oil imports, there are essentially two options: reduce consumption or switch to alternative fuels. Consumption can be reduced through efficiency measures (for example, moving toward a hybrid-vehicle fleet), changes in driving behavior (say, reducing the distance traveled or how fast the car is driven), or switching to public transportation or a bicycle. Also, switching to alternative fuels, such as ethanol made from certain types of biofuels, compressed natural gas, or electricity can be, but is not necessarily, difficult, costly, or environmentally harmful. Conventional "first-generation" ethanol in the United

States, for example, is mainly produced from corn, which has caused a huge increase in the amount of water used to generate transportation fuels. It is also not clear that energy is saved or CO_2 is reduced given the large amounts of energy required to produce corn-based ethanol (Searchinger et al. 2008). In Brazil, ethanol is produced from sugarcane, and it unambiguously results in significant reductions in CO_2, although it also requires large amounts of water to produce the sugarcane.

Policies That Help or Hinder Diffusion

There are four main types of policy that affect the global diffusion of clean and efficient energy policies: domestic manufacturing or industrial policy; technology or innovation policy; export promotion policy; and market-formation policy. The case studies and interviews indicate that these four types facilitate, catalyze, or even accelerate the diffusion of clean energy technologies worldwide.

Of course, policies can also inhibit the diffusion of cleaner energy technologies. Policies that restrict foreign access to a domestic market clearly limit the global diffusion of cleaner energy technologies. A classical example of hindrance would be the establishment of a tariff. Strange as it may seem, many tariffs exist on environmental goods in services. Such tariffs are generally low in advanced industrialized countries, but they can range from 20 to 40 percent in many developing countries (Monkelbaan 2011). Many concerns internationally were raised about "buy American" provisions in the American Recovery and Reinvestment Act, for example, which stipulated that "none of the funds appropriated or otherwise made available by the Act may be used for a project for the construction, alteration, maintenance, or repair of a public building or public work unless all the iron, steel, and manufactured goods used are produced in the United States" (Energy Efficiency and Renewable Energy 2012). Similarly, China's "indigenous innovation" policies worried many foreign firms because they believed these policies would be a distinct disadvantage in the Chinese marketplace. In the case of advanced batteries, these indigenous innovation policies caused one US firm interviewed to seek a joint venture in China so that its products would be considered "Made in China."[2] In this case, the US firm was relatively philosophical about the need to do so, noting that it would like to be close to market anyway.

Regulations can inhibit technology diffusion as well. One type of regulation cited as a barrier is when electric utilities regulators prohibit the use of certain technologies.[3] Permitting policies can also limit the diffusion of certain types of technologies. Licensing prohibitions, either self-imposed by firms or imposed by governments, can also create an obstacle to cross-border diffusion. Export controls are a clear instance of such prohibitions, and US export controls have inhibited the licensing of gas turbine technologies by Chinese firms from US firms.[4] "Dual use" technologies—those with both military and civilian applications—are often required to be reviewed by the exporting government, and so at a minimum, this review process slows down the licensing process even if it does not prohibit technology transfer in the end. A common form of technology acquisition is the purchase of a foreign firm, and this usually entails a government review as well.

Policies can conflict with one another, sending different or even contradictory signals to the market. One example of contradiction relates to the goal of reducing fuel consumption in automobiles. Some governments have established fuel-efficiency standards, but then combined these performance standards with fuel subsidies, making fuels cheaper, with the result that consumers are actually encouraged to use more fuel, not less. While fuel-efficiency standards typically reduce the fuel consumption of a product overall, there is usually a "rebound effect" in which the consumer uses the product a little bit more because it is cheaper for them to do so since the product is more efficient.

Domestic Manufacturing or Industrial Policy

Domestic manufacturing or industrial policy—hereafter simply called industrial policy—may not, at first blush, seem like a type of policy that would lead to the greater diffusion of clean energy technologies around the world. After all, domestic industrial policy is aimed at strengthening domestic firms, and many of the typical measures for doing so involve protecting domestic firms from international competition through the imposition of tariffs or other trade barriers, among other mechanisms. But domestic manufacturing seems to be essential to global economic integration. As one US government official explained, "You cannot export products you don't manufacture."[5] The establishment of new firms with the support of industrial policy can also counter oligopolistic industrial structures and therefore reduce costs for technology consumers. In China, the high costs of foreign technologies have motivated the government to support Chinese firms to develop their own technologies in some sectors.

Essential to the development of domestic industry is the availability of capital. If firms cannot borrow, they usually cannot expand. Capital is also required if the firm wishes to invest in R&D, or acquire technology from another firm through a license, investment in a joint venture, or outright acquisition of the other firm. Access to capital for clean energy firms has been a particular challenge in most countries because the firms are usually new and lacking an established credit record. Their products or services are deemed risky, especially in cases where the market is ill defined, unstable, opaque, or unpredictable. This problem of risk is one of the main reasons why market-formation policies are required (as explained below). In cases where technologies, products, or services are deemed risky, the cost of capital either rises or simply is unavailable. In general, Chinese clean energy firms have enjoyed virtually unlimited amounts of capital. Their ability to obtain loans at low interest rates allows them to acquire land and build factories cheaply. They can purchase manufacturing equipment and pay for technology licenses, along with the services of foreign experts who can teach them, as needed, how to assemble a manufacturing line or operate equipment. They can even buy foreign firms or develop technology partnerships with research institutes and universities, domestic and foreign.

The Chinese wind and solar cases are excellent illustrations of how access to finance can be a precondition for success.[6] In other situations, however, access to finance was either not available in an earlier period (that is, coal gasification) or not the core problem (such as advanced batteries and gas turbines). Firms interviewed in the United States cited a lack of access to capital as a primary barrier to their export competitiveness. Even when a market could be defined abroad, some US firms were unable to secure the financing they would need to serve that market. In contrast, Chinese firms targeting renewable energy markets in other countries were able to get financing domestically. The main US approach to the provision of finance during the last decade was to offer loan guarantees to clean energy firms. The conceptual idea certainly was good. By providing government backing to commercial loans, the cost of capital could be reduced at low risk to the government, and capital would then flow to emerging clean energy firms. The demand for the loans was high, but the government proved conservative in its desire to avoid placing any risk on taxpayers. The total amount of loan guarantees provided was $34 billion as of September 2012 (DOE 2012b). Even more helpful to firms were the cash payments in lieu of investment tax credits in the

US Section 1603 program of the American Reinvestment and Recovery Act (ARRA).[7]

The primary downside to the Chinese approach is overinvestment. When an emerging industry shows signs of success in China, copycatting becomes rampant and a new factory sprouts up in every province. Every governor and local mayor wants to invest, and they provide incentives to local entrepreneurs to set up local factories, which create local jobs and diversify the regional production base. Sometimes this works, but frequently it leads to ruthless competition among Chinese firms that either oversupply the market, causing prices to crash, or undercut each other in an effort to put their competitors out of business. While this phenomenon is good in the short term for consumers inside and outside China because it brings prices down, it ultimately harms the industry. As Chinese firms compete with each other so fiercely, they become focused on the short term. They fail to make investments in innovation or think about the medium to long term, so focused are they on delivering profits for the quarter to their investors. After a while, the cycle plays out with the weaker firms failing, jobs lost, and poor investments written off. The central government has railed against this overinvestment on the part of local governments, but it has not solved the problem of how to make capital available to entrepreneurs without causing overinvestment. In other words, the Chinese government does not have an adequate "reciprocal control" mechanism in place for clean energy where government support is made conditional on performance (Amsden 2001). Still, if the choice is between taking the risk of providing low-interest loans in the public interest or not, the evidence strongly suggests that the government had better make the loans.

The second, third, and fourth components of industrial policy are closely intertwined: scaling up in the domestic market, achieving commercial demonstration, and national branding. Firms need to be able to scale-up somewhere, and they prefer to do this in their home market. They are most comfortable working in their own culture, with their own workers, and often require some government support to get started. They expect to learn as they grow, and hope for some national forgiveness for mistakes made and lessons learned, especially if they are managing to provide jobs to local workers. It is unnatural for these firms to look abroad for that government support, but sometimes they are forced to do so, as the cases of Evergreen Solar and A123 showed. If these firms are lucky, they are able to successfully demonstrate their products at home, prove their commercial viability to banks and potential customers, and then quickly achieve economies of scale. Many firms emphasized the importance of demonstration. You have to be careful when you introduce a technology for the first time because you may never get a second chance, but if you are successful, it is extremely helpful not only to drive sales at home but also begin exporting.[8] If foreign buyers can see that the technology is working in one place, it greatly increases their confidence that the technology will work in a different context.

Achieving scale at home can allow a firm to drive down costs (though this is not always the case, as explained in chapter 6). It can create local jobs and allow a firm to be close to market (if the market exists; see the section below on market-formation policy). If many firms in one sector are able to scale up at approximately the same time, in relative spatial proximity, agglomeration effects can occur due to backward and forward linkages in the economy (Krugman 1997). This agglomeration effect is clearly evident in China where firms in the clean energy sector are able to respond to changes in the marketplace almost immediately due to the proximity of suppliers. Foreign firms have moved to China as well to be closer to the rapidly emerging clean energy market for wind, solar, and advanced batteries especially. An emerging agglomeration effect seems to be occurring in certain regions of the United States, such as Ohio, where the rust belt appears to be greening, and the American West, where market-formation policies, specifically the renewable portfolio standards of Colorado and Texas, have created such strong local demand that firms from around the world are colocating close to the market. An unanticipated agglomeration effect took place in Germany after the formation of the feed-in tariffs there. While this effect was not planned for or designed, it is one of the happiest outcomes of the German Renewable Energy Law from the point of view of German firms, government officials, and workers employed by the industry. The initial goal of the law was to simply accelerate the deployment of clean energy technologies. There were no industries lobbying for its establishment; rather, a political party—the Greens—wanted to reduce pollution and shift away from nuclear energy. Little did the Greens know that their initiative would lead to the creation of more than 367,000 clean energy jobs twenty years later (Federal Ministry for the Environment, Nature Conservation, and Nuclear Safety 2011).[9]

If industrial policy is successful, a national brand is established. Many of the interviewees from US firms wondered aloud about whatever happened to "Made in America"—still a strong brand internationally.[10] Although GE is well known for its clean energy products and launched a worldwide *ecoimagination* branding campaign, most of the smaller US firms do not yet benefit from a Made in America clean energy brand. The United States is no longer known for its clean energy products the way it was in the 1970s. While Made in China has become ubiquitous and is often associated with inferior quality, in the clean energy sector, Made in China is more approvingly viewed as "more affordable." US firms are licensing Chinese coal gasifiers. US solar distribution and installation companies whose businesses were growing rapidly until the US government imposed steep tariffs on Chinese panels are importing Chinese solar PV modules. The cheaper Chinese solar PV modules were enabling off-grid solar systems to be deployed in rural areas in many developing countries because their cost was finally in reach for poor villages without access to electricity. "Made in Germany" has become synonymous in the clean energy space with premium quality and high technology. "Made in Japan" is equivalent with clean and efficient automobiles due to Japanese leadership in this sector.

Finally, standardization is helpful. A country can establish technical and regulatory standards that make agglomeration easier. Standardization smooths the backward and forward linkages within an economy that can lead to economies of scale. A number of interviewees noted that China lags both the United States and Germany in its standardization, and many more argued that if countries could harmonize their standards, the *global* diffusion of cleaner technologies would be greatly facilitated.

Technology or Innovation Policy

Technology policy is traditionally understood to be government support for R&D, but actually policy is required for all aspects of an innovation system. There are many feedbacks among the so-called stages of innovation (such as RD&D, early deployment, and widespread diffusion), and governments must encourage the information feedbacks among stages, play a coordination role, provide funding for different functions, encourage entrepreneurship, foster spillovers, and establish the market-formation policies. The technology leadership enjoyed by the Chinese today in coal gasification would never have occurred without the initial licenses purchased by the Chinese government, sustained government investments in R&D on the part of MOST and CAS, and support for demonstration of the new Chinese gasifiers by MOST and the NDRC. The Chinese government has not always been consistent in its support for the energy innovation system as a whole, but it seems to be recognizing the importance of a systemic approach today. While it provided support to the early Chinese manufacturers of solar panels, it did not originally invest large sums of money in solar R&D, nor did it establish market-formation polices as early as European governments did. Now it is doing both. MOST has supported the development of cleaner vehicle technologies since the tenth five-year plan, and it tried to establish market-formation policies in the form of purchase incentives for EVs, but neither are yet sufficient. Now that there is fairly strong market demand for gas turbines in China, the government finally ramped up its R&D support for gas turbine technology for the twelfth five-year plan.

Export Promotion Policy

The RE4 Advisory Committee (2012) to President Obama's renewable energy and energy-efficiency export initiative has noted that a "lack of access to capital is a primary barrier to [US] export competitiveness." Export promotion is thus buttressed by industrial policy. Some countries have large Ex-Im or development banks, and these institutions greatly facilitate the cross-border movement of clean energy technology. The Ex-Im Bank of the United States is relatively small by comparison, and many US firms interviewed expressed fervent wishes that it could be more robust, flexible, and competitive with other countries' Ex-Im banks. Some of the Ex-Im Bank's regulations impose high costs on US producers as well, such as the requirement to use American-flagged ships for shipment, which are often 30 to 50 percent more expensive than those of other nationalities. Still, the US Ex-Im Bank and the Overseas Private Investment Corporation appear to have made significant efforts to boost their support for renewable energy and energy efficiency. As of 2011, the Ex-Im Bank had provided $700 million in financing to foreign buyers of US technology, almost double the previous year, and the Overseas Private Investment Corporation supplied $520 million in risk insurance and loans. The US Commerce Department engaged in many new trade missions for clean energy as well (RE4 2011).

Other countries have implemented other export promotion strategies. Germany launched a renewable energy export initiative that includes the creation of catalogs and trade missions as well as support for physical demonstrations of clean energy technologies in other countries, such as the installation of German solar panels on a local school.[11] They have found that such demonstrations are helpful in convincing local governments that the technologies work as advertised. Other countries tie their development aid to the use of their clean energy technologies, and Japan is perhaps most famous for using this technique. China recently established a multimillion-dollar aid initiative for Africa that will deploy Chinese solar panels to forty African nations (*China Daily* 2012a).

Market-Formation Policy

Market-formation policy is not the same as the creation of niche markets. A theoretical discussion of the difference is articulated in chapter 7, but suffice it to say here that clean and efficient energy technologies require broad and sustained markets due to the pervasive market externalities and distortions they face. Policy is required to create demand for cleaner and more efficient technologies. This policy must be stable, predictable, transparent, and medium to long term. It probably does not need to be permanent, however, and should be designed from the start to be able to respond to changes in technology costs and the marketplace. Subsidies are politically hard to remove once they are created unless it is clear from the start that the subsidies will phase down as market conditions change. Senator Lisa Murkowski (2011) of Alaska recognized this problem when she stated during a hearing that "we start with the subsidies and then people become very attached to them. In representing our constituents that have gotten attached to them it's difficult to undo them. We have subsidies in place that we probably don't even know are still out there. We're paying for them." A more academic version of this insight found that rent-seeking behavior is inevitable whenever the government intervenes in the marketplace (Krueger 1974). The Spanish government discovered it had created a problem with its politically popular but government-revenue-sapping feed-in tariffs for renewable energy. Debts worth twenty-eight billion euros had accumulated as of 2013 due to Spain's feed-in tariffs and other subsidies (Johnson 2013). Likewise, Germany discovered that its own feed-in tariffs had become corporate welfare when firms were pocketing the growing difference between the falling costs of renewable energy technologies and the guaranteed tariff (Weismann 2012). Once a cleaner energy technology can compete in the marketplace on its own, the rationale for government intervention is greatly diminished, if not eliminated, because there will always be externalities related to the burning of fossil fuels.

Besides being stable, predictable, transparent, and medium term in nature, other desirable characteristics of market-formation policies are that they must be aligned, reward performance and not capacity, and impose penalties for noncompliance.[12] Aligned means that the policies should not send conflicting signals to the marketplace. The policies should reward performance—that is, the actual generation and use of clean electricity—not the construction of the wind turbine (a problem Germany is coping with as offshore wind developments in the North Sea cannot be connected to the German grid) (Casey 2013). If market-formation policies are being implemented as part of an industrial policy strategy, they cannot protect the domestic market for too long because then the domestic industry becomes complacent, unable to compete with foreign firms, and unable to export to markets abroad.

Analysis of the case studies in this book led to the finding of the importance of market-formation policies. The birth of the Chinese solar industry was inspired by the burgeoning demand for solar PV in Europe, Japan, and the United States, and that demand was created by feed-in tariffs in Europe, government-sponsored deployment programs like the Subsidy Program for Residential PV Systems in Japan, and renewable portfolio standards in the United States. While the Chinese government did not initially form a domestic market for solar PV, it soon recognized the need for it, and established the Golden Sun demonstration program and then a feed-in tariff for solar PV. Within the Chinese market, a strong demand has always existed for synthetic natural gas due to China's lack of domestic gas resources, and therefore a natural market demand for coal gasification has existed for decades. This natural market demand and obvious national need inspired the Chinese government to support RD&D in coal gasification technologies. On the other hand, the lack of natural gas resources in China convinced many government leaders that there was no reason to support R&D in gas turbine technology since they thought there would never be demand for gas turbines in China. Now they rue the day they bet against natural gas turbines because they have realized that with domestic gas turbines they could more affordably use gas in industrial applications as well as pursue IGCC as an electricity strategy.

In other words, the lack of market demand in China inhibited Chinese firms from developing and deploying gas turbine technologies, and they are now decades behind their US, European, and Japanese competitors. Likewise, market demand for advanced automotive batteries has thus far been weak. Governments around the world have not yet established robust market-formation policies for highly efficient vehicles, although in some places they are beginning to get close. Meanwhile, the costs of these batteries have remained high due to their technological complexity. Some combination of technological push and market pull along with more careful management of the innovation system for advanced batteries overall is still required (Nemet 2009). While the case studies are persuasive about the importance of market-formation policies, two other methods were employed to test the proposition.

To provide a quantitative snapshot of the impact of national and subnational market-formation policies on the cross-border movement of clean energy technologies, the change in the volume of trade in clean energy technologies was compared with the cumulative number of national and subnational market-formation policies per year. Total import and export data for solar PV modules, wind energy technologies, gas turbines, and lithium batteries globally were obtained through the United Nations' Comtrade (2012). Wind energy technologies were included even though they were not the subject of one of the case studies because many of the market-formation policies for clean energy are broadly aimed at incentivizing the use of renewable energy technologies. Data on coal gasification and gas turbine technologies could not be cleanly extracted from Comtrade so they were omitted. Many other end-use technologies as well as cleaner supply technologies like biomass, geothermal, and nuclear could also be included in such an analysis. They too were omitted because they were not examined in the case studies.

Major market-formation policies at the national and subnational level were identified in countries around the world from 1970 to 2012 using dozens of sources, and a timeline of these policies, along with references, is contained in appendix D. The designation of a policy as "major" was a matter of judgment because it was difficult to determine with any specificity the likely market impact of each market-formation policy. To construct the list, initially the implementation of all renewable portfolio standards or quota policies and feed-in tariff policies was recorded based on Ren21 (2012) reports as well as the IEA/International Renewable Energy Agency (IRENA) database. Additional market-formation policies were identified in the IEA/IRENA (2012) database beyond renewable portfolio standards/quota and feed-in tariffs by searching by jurisdiction. No supranational policies were included. Information about policies that were obviously missing from these two primary sources was supplemented through a wide variety of sources all referenced at the end of appendix D. No push policies were included, meaning that all RD&D policies and investments were excluded from the list, since the main purpose was to assess the global totality of national and subnational market-formation policies.

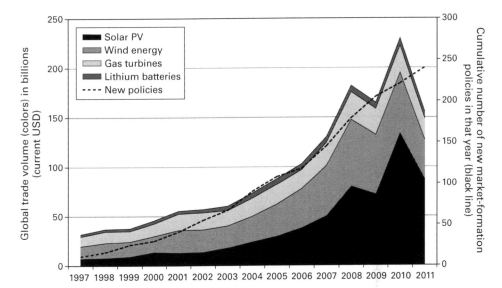

Figure 4.1
National and subnational market-formation policies and international trade in clean energy technologies

The volume of world trade was then compared with the cumulative number of major new market-formation policies in each year. The results for 1997–2011 are presented in figure 4.1. The chart starts in 1997 because between 1970 and 1997 there were ten or fewer policies per year, but policies for the prior years are documented in appendix D. The global uptick in the number of national and subnational policies coincides with the establishment of the Kyoto Protocol in 1997, which is interesting because the protocol only established binding emission-reduction commitments for a subset of industrialized countries (the US Senate refused to ratify it), and there were no such commitments for industrializing countries even though many of them went on to create market-formation policies. The Kyoto Protocol therefore appears to have catalyzed policy for clean energy market formation. The subsequent Conference of Parties to the UNFCCC in Copenhagen in 2009 appears to have sustained the growing trend in national market-formation policy, but disappointment with the outcome may have led to the sharp decline afterward, although time will tell. The 2009–2011 period also coincides with the global recession so not all the blame can be placed on the disastrous negotiations in Copenhagen.

As figure 4.1 shows, international trade in clean and efficient energy technologies grew dramatically during the 2000s, especially for solar and wind technologies. The sharp rise in solar PV exports and imports is particularly remarkable. Since 2002, the growth in the volume of trade for both solar and wind appears to lag by a year or two the enactment of new policies around the world that have the effect of stimulating markets in renewable energy. This outcome is probably due to the fact that after new laws are enacted or regulations promulgated, most require time for implementation, after which time producers then can respond to the new market incentives. The passage of China's renewable energy law, launch of the EU Emissions Trading System as well as Norway's domestic emission trading regime, reestablishment of the US production tax credit for renewable energies, establishment of feed-in tariffs in places as widely spread as Ecuador, Ireland, and Uttar Pradesh (India), and renewable portfolio standards or quota regimes in three US states (including the District of Columbia)—all in 2005—appear to mark the beginning of significant growth in trade volume. California's renewable portfolio standard, the reestablishment of the Production Tax Credit in the United States, and the United Kingdom's Renewables Obligation—all in 2002—together with smaller-scale policies in Algeria, Australia, Austria, Brazil, Canada, the Czech Republic, France, Indonesia, Israel, Lithuania, and Wallonia (Belgium) seemed to kick off the tremendous growth in global trade for the decade.

Figure 4.1 illustrates the trends in market-formation policies, and how they compare with trends in global trade for cleaner energy technologies. Figure 4.2 presents a scatter plot resulting from a simple bivariate regression between these two variables with the cumulative number of new market-formation policies on the x-axis and the total volume of trade in the four cleaner energy technologies on the y-axis. As the plot indicates, there appears to be a correlation between the cumulative number of market-formation policies and global trade given that R^2 is 89.9 percent. Recognizing that this is solely a bivariate regression, and could suffer from omitted variable bias, multicollinearity, and other deficiencies, it is nonetheless encouraging that the plot reinforces the findings from the case studies.[13]

China accounts for much of the growth in world trade in clean energy. The total volume of trade in clean energy is expanding rapidly—more than twice as fast as manufactured goods overall. For these four clean energy technologies alone, the volume of exports and imports grew 259 percent between 2000 and 2010, compared with 118 percent for the

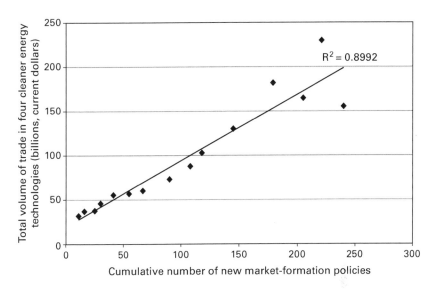

Figure 4.2
Relationship between cumulative market-formation policies and the international trade in clean energy technologies

exports and imports of the total manufactured goods globally.[14] Chinese solar PV exports were only worth ninety-three million dollars in 1997, accounting for just 3 percent of global PV exports, but by 2011, Chinese PV exports had soared to twenty-seven billion dollars, accounting for 45 percent of the global exports. China is far from the only developing country contributing to the global marketplace in cleaner energy technologies. China together with Indonesia, Malaysia, Mexico, Poland, and Hungary, for example, accounts for 20 percent of the global LED exports.

Summary

Government policy strongly affects the deployment of cleaner and more efficient energy technologies around the world, especially domestic manufacturing or industrial policy, technology or innovation policy, export promotion policy, and market-formation policy. Within these categories, the provision of access to affordable finance and policies that create demand for the use of cleaner technologies appear to be the most important. Market-formation policies at the subnational and national level are strongly correlated with the expansion in global trade for clean energy technologies. Many different market-formation policies have been experimented with around the world, and while feed-in tariffs and renewable portfolio standards have thus far been the most widely used, other policies including carbon taxes and performance standards are also effective. To the extent that market-formation policies are harmonized across countries, the demand would become more uniform, which would undoubtedly be helpful for producers.

5

No Risk, No Reward

A Chinese CEO of an EV company interviewed for this book had inti-
mate knowledge of the landscape of patents in his domain. In his view,
international patent protection is "very strong," in HEVs so there is
almost "no way to get around the patents if you don't want to steal their
technology."[1] To him, it is obvious that trying to do so would be illegal,
and thus he believes that the Chinese have greater potential for success
in pure EVs since the foreign patenting has been less comprehensive.
According to him, in general, with the exception of the HEV patents,
Chinese entrepreneurs can readily license foreign technology (although
they might be prohibited from exporting) and they should rely more
heavily on indigenous development. But as this experienced CEO frankly
said during an interview, "You know I have been studying the EV situation for many years now, and in China, the most important barrier is
price" (ibid.). In other words, access to intellectual property is a lesser
concern relative to weak market demand for EVs in China and around
the world given their high price.

Dave Vieau of A123 Systems, an advanced-battery firm, had a different perspective but arrived at the same basic conclusion.[2] When A123
began exploring opportunities in China, one critical element was how to
manage the risk of bringing A123's intellectual property to China while
maximizing the potential reward. Vieau believed there were two main
ways a company could lose its technology in China: either rank-and-file
employees get to know the production process well and start their own
business, or else the Chinese partners or high-level firm leadership strike
out on their own. Initially, to mitigate the first risk, A123 decided to
pursue what Vieau called a "foundry model," where production is separated from design. A123 in fact not only separated design from production but also separated various parts of the production process and sited
them in different locations. For a variety of reasons, this model did not

work, and the firm turned to a different model based more on partnership and shared economic interests. It started to realize there was a large potential market in China based on developments in Chinese policy and began to file patents there. Yet for all the business challenges related to protecting intellectual property in China, Vieau noted that China's cost advantages and favorable finance environment trumped the risks related to intellectual property. He was convinced the Chinese government would make a long-term commitment to fostering demand for advanced automotive technologies, which is needed given their currently high costs.

As illustrated by these two viewpoints, there are two main concerns about technology transfer and intellectual property, deriving from opposite perspectives. First, technology owners and suppliers are often fearful of transferring technology and knowledge to countries with intellectual property regimes that are perceived as weak or ineffective. Technology owners may fear infringement, misappropriation, lack of remedy through the courts, and loss of competitiveness if recipient firms are able to catch up through unfair means. These concerns have been well documented in interviews for this book, the scholarly literature, and the popular media. Indeed, China epitomizes the concerns many foreigners express with its reputation as a place where it is risky to bring intellectual property, even if this reputation is no longer fully warranted, or if these concerns relate to certain industries more than others.

From an aspiring firm or country's perspective, potential purchasers of technology worry about barriers imposed by foreign firms or intellectual property regimes to limit access to advanced technology. Previous studies have found limited evidence of barriers to the acquisition of some cleaner technologies (specifically alternatives to ozone-depleting substances, some technologies used in HEVs, wind turbines, and gas turbines), including prohibitively high licensing fees, refusals by foreign companies that are unwilling to license their technology at any price for fear of generating competitors, or "excessive" patenting done to crowd out potential competitors (Gallagher 2006a; Lewis 2007; Watal 2007; Watson et al. 2010). In the international climate negotiations, many developing countries express concerns about access to intellectual property, and some (specifically India and Bolivia) have even called for compulsory licensing of cleaner energy technologies (Maskus and Okediji 2010). It should be clearly stated that the Chinese government has never called for the compulsory licensing of energy technologies, though it often asserts the need for more technology transfer and enhanced financing for climate mitigation.

Intellectual property plays an important role as an incentive to technology development and transfer (Abdel-Latif 2013). This chapter explores the extent to which intellectual property regimes incentivize *and* inhibit the international transfer of cleaner energy technologies. The main finding is that cleaner technologies diffuse across borders despite intellectual property barriers if the market incentive is sufficiently strong. Why intellectual property is not a bigger barrier is not clear, but a set of hypotheses are offered that can be tested in the future with additional empirical evidence. There are four main intellectual-property-related barriers to the cross-border transfer of cleaner and more efficient energy technologies. First, if intellectual property protections are too strict and comprehensive, they can unnecessarily inhibit spillovers and learning. Second, if host country intellectual property protections are too weak, they can inhibit inflows of technology due to foreign concerns about infringement. Third, intellectual property acquirers can be incapable, unknowledgeable, and unable to effectively acquire and absorb technologies. Fourth, intellectual property holders can be too monopolistic, stingy, or risk averse, and may be unwilling to sell or share intellectual property at all. In this chapter, these four barriers are examined in light of the case studies on gas turbines, batteries, solar PV, and coal gasification.

Specifically, the questions that will be answered in the paragraphs below are: Is there evidence of widespread or systematic infringement in clean energy technologies in China? Second, is there evidence of unreasonable withholding of certain cleaner technologies by foreign firms or countries? Is the cross-border flow of clean and efficient technologies growing or diminishing based on what we know about the movement of these technologies through patent data in China?

The chapter begins with a synthesis of the state of knowledge about intellectual property and international technology transfer in general. Then, Chinese policy for its emerging intellectual property regime is explained. In the third section, the evidence in China and elsewhere is examined, based on the case studies, court cases in China, peer-reviewed literature, news articles, and other interviews. Next, an analysis of trends in patenting in clean and efficient energy technologies in China is presented, documenting the explosion of patenting in this space across the globe and especially in China. Finally, overall findings about intellectual property and technology transfer are discussed.

Intellectual Property and Technology Transfer

The scholarly literature is inconclusive about the degree to which intellectual property regimes affect technology transfer in general. As summarized by Keith Maskus (2000), "By providing additional certainty about the enforceability of contracts, intellectual property rights (IPRs) could encourage firms to trade technology across borders through making costly investments in FDI and licensing. But, by raising the cost of imitation, IPRs might limit international diffusion through unauthorized means." Many studies to date find that a strong domestic intellectual property protection regime has positive effects on the country's imports, licensing, and inward foreign direct investments (Copenhagen Economics 2009; Branstetter et al. 2007; Branstetter, Fisman, and Foley 2006; Park and Lippoldt 2005). Other studies have shown that patent protection is less important than issues like retaining market position, achieving more efficient methods of production, and gaining short-term advantages for corporations making R&D decisions, except in the pharmaceutical and certain other industries (Scherer et al. 1959; Taylor and Silberston 1973). Historically, countries' intellectual property rights regimes coevolve with their economies, and the relative merits of different intellectual property policies and regimes change during the stages of economic development (Odagiri et al. 2010). The degree of intellectual property protection needed for certain industries (for example, chemicals, pharmaceuticals, and entertainment) is quite different from that for other industries because it is much easier to copy a DVD or replicate a chemical formula than it is to reverse engineer a complex technological system like a gas turbine.

A number of studies on cleaner energy technologies have concluded that the effect of national intellectual property regimes is hard to evaluate, but probably not as large an inhibitor to the technology transfer of cleaner energy technologies as conventional wisdom would suggest (Gallagher 2006a; Watson et al. 2010; Copenhagen Economics 2009; Ockwell et al. 2010; UNEP, EPO, and ICTSD 2010). This area is clearly deserving of additional empirical research, which this chapter will provide.

With respect to cleaner energy technologies, there is prior evidence that while acquiring firms or countries may have access to certain classes of technologies, they may not be able to acquire cutting-edge technologies (Ockwell et al. 2010). These "n-1" or "n-2" technologies may be all that firms are willing to sell, or the acquiring entities may be content to import a slightly older generation of technology.[3] This latter situation was coined as the "good enough phenomenon" in an earlier study of the development of the Chinese automobile industry (Gallagher 2006a). There is an obvious reason why the acquiring firms may be comfortable with not having the latest generation of technology: cost. If the performance is nearly as good at a considerably cheaper price, then it is unsurprising that older technologies are often purchased. Also, older generations are proven technologies and therefore less risky to buy.

A newer debate emerged in the 2000s about the rate and effectiveness of Chinese domestic or "indigenous" innovation in clean energy technologies. One theory, following William Alford (1995), posits that as Chinese clean tech firms catch up with their Western counterparts, Chinese government policy will prefer investments in domestic innovation over technology purchases from foreigners (Lema and Lema 2010).[4] The multitude of methods in technology acquisition that can be observed in the Chinese cases, however, could also be interpreted as evidence of a new globalization of the development and deployment of clean energy technologies. In other words, due to globalization, clean technology options abound worldwide, and acquirers have no need to use illicit methods to get technology.

China's Intellectual Property Regime for Cleaner Energy

A 2009 Chatham House report found that companies and research institutions from the United States, Japan, and Germany are the clear leaders in energy innovation as measured by patents granted. The authors argued that China lagged far behind these other countries based on the fact that China had no companies or organizations in the top ten positions in any of the sectors and subsectors analyzed for patent ownership. Yet overall, China was found to have the fourth-largest number of clean energy patents by country of origin (Lee, Illiev, and Preston 2009). In this section, China's intellectual property regime as it relates to clean energy is explained. China's first patent law was established in 1984, at the beginning of China's opening, and was subsequently amended three times, in 1992, 2000, and 2009. With China's entry into the WTO in 2001, it became a member of the Trade-Related International Property Rights (TRIPS) agreement. The 1990s and early 2000s were a period when the China's intellectual property regime took shape in general, and these years were marked by bureaucratic disputes and conflicting internal views about what China's strategy should be regarding intellectual property (Mertha 2005).

In a sharp departure from the evolutionary period, the Chinese government is now firmly emphasizing the establishment of a robust intellectual property regime in China. Wen Jiabao reportedly stated in 2004, "The future competition in the world is in intellectual property" (Cohen 2010). The Chinese government got serious during the late 2000s about creating a robust intellectual property regime. A major intellectual property strategy was announced in 2008, a standing office under the State Council was established to work on intellectual property protection issues, and in 2010, a National Patent Development Strategy (2011–2020) was announced. As of 2010, China had surpassed both Europe and South Korea to become the world's third most active patenting country after the United States and Japan. The intellectual property court system is expanding rapidly in China as well, and there was more intellectual property litigation in China in 2010 than in the United States. Copyright cases accounted for 58 percent of the total, trademark for 20 percent, and patent 13 percent, and most of these cases were Chinese suing other Chinese (Suttmeier and Yao 2011).

The Chinese government now places particular emphasis on the filing of patents. In the twelfth five-year plan, the government set a goal of granting 3.3 patents for every ten thousand people by 2015, which would double the number granted in 2010. This is the first time a patent goal has been included in a five-year plan (Casey and Koleski 2011). China is increasing the number of patent examiners with a goal of employing nine thousand by 2015.

Additionally, many, if not all, provincial governments subsidize patent filing fees, and local governments often cover the rest of the application cost and offer bonuses if patents are granted. Provincial governments also subsidize foreign patent applications. In 2009, the central government announced it would also begin to subsidize foreign patent applications up to a hundred thousand RMB per application (*People's Daily* 2009). Shanghai created the first patent subsidy program in the late 1990s, and other provinces quickly followed to the point that by 2007, twenty-nine out of thirty provinces in China had established patent subsidy programs (X. Li 2012). Hunan Province, for example, has been providing subsidies to "local enterprises, institutions, and individuals" since 2004. Hunan offers a subsidy of three thousand RMB for each invention patent obtained, and four hundred RMB for each utility model or design patent. International patents are rewarded with ten thousand RMB. As of 2011, Guangzhou's Intellectual Property Office was reviewing about three thousand applications for patent subsidies per month, and had awarded subsidies to 31,400 filers for a total expenditure that year of fifty-one million RMB (Anonymous 2011a, 2011b). Thomson Reuters and PriceWaterhouseCoopers report that government agencies also occasionally provide patent holders with tax, tenure, and residence permit benefits (Zhou and Stembridge 2008; PriceWaterhouseCoopers China Pharmaceutical Team 2009).

It is unclear how much Chinese patent data are inflated as a result of these incentives. Another concern is whether or not the incentives affected patent quality. In an analysis of the recent surge of Chinese patenting in general, Xibao Li finds that provincial-level patent subsidies indeed induced growth in patent filings.[5] To determine whether or not the surge led to a decline in the quality of patent applications, Li examines whether or not the applications-to-acceptances ratio dropped, which would be expected assuming that the criteria used to assess the patent applications did not change. He finds that the grant ratio actually increases between 2000 and 2004, which was the period of the general surge, indicating that to the contrary, there may have been an increase in the quality of patent applications. Li concludes that patent subsidies do not necessarily cause patenting "bubbles" so long as the examination process remains unaffected by the increased number and complexity of patenting (X. Li 2012). No evidence exists that the examination process has become more lax, and if anything, the rigor is increasing. As already discussed, China's SIPO is training many new patent examiners as well, so it ought to be able to handle increased applications (Harvey 2011).

There is no question that enforcement of intellectual property laws is a tremendous challenge in China, as it is in relation to many other types of policies promulgated by the central government. Certain sectors are much more notorious for infringement in China, especially media, software, and pharmaceuticals, where it is relatively easy to reproduce products.

The Evidence

Existing evidence in the scholarly literature for and against the claims made about the role of intellectual property in spurring or hindering the transfer of cleaner energy technologies across borders is slight. In this section, new evidence is presented based on court cases, cases from the WTO dispute settlement mechanism, the news media, and the case studies presented in chapter 3 about the extent to which infringement or withholding of cleaner energy technologies is occurring in China.

Infringement

When most foreign companies consider interacting with China, either as a licensor, joint venture partner, trading partner, or otherwise, their first instinct is either opportunity or fear, or some combination thereof. The opportunity springs from the large potential market, and the fear from the widespread perception, well founded or not, that if you go to China, you will lose your shirt.

A major review by the US International Trade Commission found that the copyright industries, including music, movies, software, and publishing, are indeed subject to "substantial" infringement in China, and that many US firms in these industries consider trademark counterfeiting to be one of their most serious problems. Research for this book did not find similar widespread infringement of clean energy patents, although US firms interviewed for the study had mixed opinions on whether there is an antiforeign bias in the issuance or enforcement of patents in China. Some firms have asserted that they have found it more difficult to obtain patents in sectors that the Chinese consider of strategic importance, such as pharmaceuticals, renewable energy, and biotechnology (Hammer and Linton 2011). None of the clean energy firms interviewed for this book, however, disclosed any problems obtaining patents, nor did any foreign firm experience any firm-threatening infringement.

Two foreign clean energy companies interviewed for this study disclosed problems that emerged with rogue employees.[6] The first company had actually had two cases. In the first, which was actually *not* a dispute over an energy technology, the firm was able to obtain an injunction from the State Industry and Commerce Commission and then filed a civil damage suit, which rendered a decision in the foreign firms' favor, including a substantial award, which was paid in two installments. This company's other case was related to energy, but because it was still pending, the firm could not discuss it. The other foreign firm discovered that an employee had set up a parallel company, and the Chinese courts allowed it to raid the company to determine that some of its patents had been compromised. As of the time of this writing, this firm was still pursuing the case. The livelihood of the foreign firm was not compromised in either of these instances, but of course they are troubling stories. A third instance of possible infringement has been widely reported in the media: the dispute between Chinese wind turbine manufacturer Sinovel and American Superconductor, discussed in detail below.

As a researcher, I had assumed at the outset that there would be many more cases than I actually found, and planned to study this body of evidence to determine when, how, and why infringement occurred, but in dozens and dozens of interviews I only uncovered these three cases in the clean energy industry—one of which wasn't even related to energy. These two instances, though, are almost surely representative of other infringement problems in the clean and efficient energy industries. On the bright side, two of the three cases reveal that the foreign firms were adept at recognizing possible problems and managing to contain them.

Best practices identified through interviews with these and other firms included separating the production process into different physical locations, not allowing any single employee to understand the entire process, withholding key pieces—"the special sauce"—of software and know-how, filing patents in the Chinese system, and utilizing the Chinese court system. Jack Chang, senior intellectual property counsel in Asia for GE, commented that "in the past five years, we've seen significant progress in the judicial protection for IP." Some interviewees suggested that engagement with Chinese partners was the best model because through engagement, the foreign firm creates a continuing interest in its contribution. One man said, "My general philosophy is 'hug people.' If you create the right relationship and create upside for them, it will work." Many interviewees stated that they were not aware of any cases of intellectual property infringement, and one foreign businessperson went so far as to argue that there was a "myth around (the need to) protect IP."[7]

The Chinese government publishes an online list of court cases related to intellectual property infringement.[8] It is not clear how comprehensive this list is, but it provides one window on how intellectual property cases are being resolved in Chinese courts. As of April 2012, there were a total of 36,071 cases contained in the database. It is important to note that reportedly, more than 50 percent of intellectual property civil cases were settled through court mediation as of 2009 (Supreme People's Court 2009), so these cases would not appear in the database. Keyword searches related to solar PV, gas turbines, coal gasification, and advanced batteries for vehicles were entered, and a list of cases is generated in table 5.1. This list contained dozens of disputes over utility model and design patents, but these were discarded, preserving only the disputes over invention patents. Only seven cases remained, and of those cases, 100 percent involved instances of Chinese suing other Chinese. No disputes over invention patents between foreign and Chinese firms are reported. So again through this method, no large body of evidence related to invention patent disputes is uncovered for our four technologies.[9]

Table 5.1
Reported court cases related to clean energy invention patent infringement in China

	Litigants	Nationality	Date recorded[a]
Solar PV[b]	Grenadines Electronics (Xiamen) Co., Ltd. v. Sulan Lou, Zhejiang China Commodities City Group Co., the second branch of the International Trade City	Chinese versus Chinese	December 15, 2010
	Yung-Wei Xu v. Huatuo Solar Technology (Fenghua City) Co., Ltd.	Chinese versus Chinese	December 15, 2010
	Weihai Kehua Lighting Engineering Co., Ltd. v. Jilin Chaoyu Industry and Trade Co., Ltd.	Chinese versus Chinese	December 15, 2007
Advanced batteries for vehicles[c]	Lanyi Chen v. Jianbing Chen, Jianhua Chen, Liyan Chen	Chinese versus Chinese	December 15, 2007
	Inco Advanced Technology Materials (Dalian) Co., Ltd., and Hunan Kaifengnew Co., Ltd. v. Hunan Corun New Energy Co., Ltd.	Chinese versus Chinese	December 15, 2009
Coal gasification[d]	Shanxi Institute of Coal Chemistry, Chinese Academy of Sciences v. Shanxi Qinjin Coal Gas Gasification Equipment Co., Ltd.	Chinese versus Chinese	December 15, 2007
	Beijing Jia Dei Xing Ye Science and Technology Co., Ltd. v. Harbin Shiji Heat Energy Technological Development Co., Ltd., Sino Coal Dragon Harbin Gasification Co., Ltd.	Chinese versus Chinese	December 15, 2010
Gas turbines[e]	None		

Notes: All data from Chinese government, http://ipr.court.gov.cn; data downloaded April 1–10, 2012 (Chinese language only). Only invention patent disputes are listed here. The vast majority of cases are utility model and design patent disputes.
a. Date listed on the Web site.
b. Keyword search: "PV, solar power, and solar."
c. Keyword search: "batteries, vehicle, electric power."
d. Keyword search: "coal, gasification."
e. Keyword search: "gas, turbine."

The popular press has reported on two cases related to clean energy. The first case, *Wuhan Jingyuan Environmental Engineering Co., Ltd. v. Fujikasui Engineering Co., Ltd. and Huayang Electric Co., Ltd.* (between a Japanese firm and a Taiwanese power plant operator), was related to the infringement of a FGD invention patent. FGDs reduce the sulfur dioxide emitted from coal-fired power plants, of which China has thousands. Chinese firms were required by law to install FGDs in order to reduce pollution, which created a huge market for FGD technology. In this particular case, Wuhan Jingyuan had been granted a patent from SIPO, and even though Fujikasui tried to invalidate it, it was not successful since Wuhan had filed the patent first in December 1995. The Supreme People's Court found in favor of the plaintiff, ordering the two defendants to pay Wuhan Jingyuan Environmental Engineering 50.6124 million RMB (US$7.42 million) in damages, the highest damages award ever made by the Supreme People's Court (2010) for intellectual property infringement. Given that China's patent law contains a first-to-file rule, this is the example lawyers use when trying to encourage their clients to file their patents in China.

In late 2011, American Superconductor announced that it was filing three suits against China's biggest wind manufacturer, Sinovel, over copyright infringement and allegedly stolen trade secrets. In September 2011, a former American Superconductor employee confessed to selling proprietary software to Sinovel and was sentenced to one year in prison by an Austrian court. According to different accounts, at its peak, Sinovel accounted for 70 to 80 percent of American Superconductor's revenues. The US company has accused Sinovel of making unauthorized use of proprietary turbine software code and breaching its supply contracts. Sinovel has denied wrongdoing and has filed two counter lawsuits. The case is politically sensitive for many reasons. First, one of Sinovel's major investors is reportedly New Horizon Capital, cofounded by Wen Yunsong, also known as Winston Wen, the son of China's premier, Wen Jiabao. Second, Austrian investigators found a signed contract between the Serbian employee of American Superconductor and Sinovel that had been signed by Sinovel's CEO, indicating that this was not a case of a few rogue employees within Sinovel but instead had been authorized at the highest levels of the firm. American Superconductor evidently took on a great deal of risk by relying so heavily on Sinovel as a customer and perhaps it did not secure its technology in Austria as well as it should have. On the other hand, from what is known based on reporting about the Austrian court case, Sinovel appears to have planned quite deliberately to make American Superconductor irrelevant to its business in China (Riley and Vance 2012; Lappin 2011; Goossens 2012; Hook 2012). Because the case is still pending as of this writing, it was not possible to make any definitive determinations.[10]

A final source of information about possible intellectual property infringement is the WTO dispute settlement database.[11] As of July 2012, China had been engaged in eight disputes as a complainant, twenty-six cases as a respondent, and eighty-nine cases as a third party. Most of the cases where China was the complainant were filed against the United States, and none directly involved energy technologies. Of the twenty-six cases where China was the respondent, four were related to energy technologies, and one was related to intellectual property. Dispute DS362, "Measures Affecting the Protection and Enforcement of Intellectual Property Rights," occurred when the United States requested consultations about trademark counterfeiting and copyright piracy, but none of these consultations involved energy technologies directly, nor did they involve patent infringement. Two of the disputes involved export restrictions on raw materials and rare earths (DS394 and DS433, respectively), one on grants, loans, and incentives to Chinese enterprises (DS387), and one specifically on grants, funds, or awards to Chinese wind power equipment providers. In summary, none of the WTO disputes involving China so far have had to do with patent infringement in any industry, including energy.

Two main findings emerge on the infringement side from the case studies, interviews, popular press, and literature. First, foreign firms consider it a significant business challenge to protect their intellectual property in cleaner energy technologies in China. On the other hand, there are surprisingly few instances of devastating intellectual property infringement in China in the clean energy sector. This is an unexpected finding—one that stands in striking contrast to the conventional wisdom—and it therefore deserves serious reflection. Why do we not observe more evidence in China of infringement in cleaner and more efficient energy technologies? The following are a set of hypotheses that might explain this outcome, and evidence for or against each hypothesis:

Hypothesis 1   Foreign firms are reluctant to pursue court cases because they don't think they would win in a Chinese court, or that it is worth the trouble.

Numerous firm representatives interviewed expressed concern in interviews about whether or not they could win in a Chinese court, but of all those who expressed concern, *none* had tried to advance a lawsuit in Chinese courts. Two foreign firms had pursued court cases in China, and both said they thought they had received a fair hearing. Both these firms were large, multinationals with extensive experience in China and overseas litigation more generally. Of those firm representatives who did express skepticism about utilizing the Chinese court system, there was a further concern that if they did file suit against a Chinese counterpart, they would receive negative press, which could hurt their market. Some firms said the infringement they had experienced was "not worth" the trouble or cost of filing a lawsuit, and others said they didn't want to irritate or upset their relations with the Chinese government.

Hypothesis 2   Clean energy technologies are not sufficiently mature to warrant significant litigation yet.

This argument is advanced by Eric Lane (2013, 83), who states, "Patent litigation becomes a financially sound business tactic only after the products at issue are scaled, widely commercialized, and profitable. . . . Thus, the commercial, financial, and temporal realities of patent litigation limit patent enforcement in clean tech to the most mature sectors." In his book *Clean Tech Intellectual Property*, he reviews a number of disputes related to wind, LEDs, HEVs, and first-generation biofuels internationally (none of which include Chinese firms).

Hypothesis 3   Many cases don't go to court because they are arbitrated instead.

No data could be found on the percentage of clean energy infringement cases that were mediated or arbitrated, but according to a 2009 document published by the Supreme People's Court (2010) of China, more than 50 percent of "recent" intellectual property civil cases overall were settled through court mediation in the first instance. Thus, arbitration seems to be a common practice and likely explanation.

Hypothesis 4   Energy technologies are complex systems, requiring a great deal of tacit knowledge, and hence they are hard to copy. If Chinese technological capabilities are weak, then Chinese firms would not be able to copy, and there would be little deliberate infringement of intellectual property, and as such, little evidence of infringement.

Indeed, energy systems are usually complex, and this is certainly true for all four of the case studies evaluated in this book. But Chinese capabilities are quite good in at least two of our cases (coal gasification and PV), and not so good in two others (batteries and gas turbines). Using the logic above, we would then expect higher levels of infringement in coal gasification and PV, but this is not observed.

Hypothesis 5 Chinese capabilities are already quite strong, and so they do not need to infringe on foreign intellectual property, and moreover, they have strong interests in protecting their own intellectual property. This argument holds true for coal gasification and PV, although those capabilities were acquired differently. Yet Chinese capabilities are extremely weak in gas turbines and ambiguous in advanced batteries for vehicles.

Hypothesis 6 Chinese firms have increased their own understanding of intellectual property rules and requirements, which has enabled them to play within global intellectual property rules and avoid inadvertent infringement.[12]

Many Chinese firms are so focused on short-term job creation and local competition that they are not willing to put in the investment to develop strong technological capabilities in order to better absorb technologies. If they can license or otherwise purchase technology, get it into production, and make money in the short term, then they achieve their objectives. It is true that Chinese firms have had widespread success obtaining and acquiring advanced cleaner energy technologies, with a few exceptions. The lack of barriers to acquisition through mainstream means reduces the incentive on the part of the Chinese to cheat or pursue shortcuts.

In sum, if China is reputed to be the most notorious of infringers of clean energy technologies, yet we do not observe widespread infringement there *in this sector*, then at the global scale, infringement of cleaner energy technologies is likely to be small. Alternatively, since evidence of intellectual property infringement in cleaner energy technology also exists outside China, we could hypothesize that there is nothing particularly special about China in terms of its intellectual property environment for cleaner energy.

Withholding
Little evidence could be found that foreign firms have withheld clean energy technologies from Chinese firms, but there are two technologies where some evidence of foreign withholding does exist: hybrid vehicle technologies and gas turbines. In both cases, there are few foreign producers of these technologies so the industry structure is oligopolistic. Many patents have been filed around these technological

systems as well. In the case of gas turbines, there are also formal export controls.

The case study on advanced batteries for vehicles did not actually reveal any withholding on the part of foreign firms, but the issue is with the broader automotive systems for which the batteries are intended. In the case of HEV technologies, numerous interviewees noted that Toyota had refused to license HEV technologies to Chinese entities. Toyota has also obtained many patents related to its hybrid vehicle system, and Chinese experts do not believe they can find a viable way around the patents through invention. The electronic controls that manage the interaction between conventional and electric engines are believed to be the main area where patenting is pervasive. According to Chinese interviewees, Toyota was willing to license hybrid vehicle technologies to a US automaker, Ford Motor Company, and the Chinese experts believe this is because Ford's technological capabilities are strong so it was just a matter of time before Ford would find a way and Toyota could profit from its license in the intervening time. Toyota has also not been willing to manufacture hybrid vehicles in China without the importation of complete knockdown kits, which means that it is only willing to assemble Japanese-made parts in China. Partially as a result of the Chinese pessimism about being able to acquire HEV technology, the overall industrial strategy has shifted to striving for pure EVs or fuel cell vehicles.

The case of gas turbines is rather different. Older generations of gas turbines have been licensed to Chinese manufacturers, but none of the foreign manufacturers are willing to license the latest (and not coincidentally most energy-efficient) gas turbines. GE said it was not willing to sell its most efficient gas turbine to China as a final product, but Siemens stated it was willing to do so if any Chinese buyer wanted to buy it, implying it would charge a high price.

Patent Analysis

Another way to measure foreign countries' willingness to transfer clean energy technology to China is to analyze the trends in patenting there.[13] If foreign firms are filing invention patents in China, especially at an increasing rate, then it would appear that they have a growing degree of confidence in Chinese patent protection. Foreign firms would only go to the trouble of filing patents applications in countries where they anticipate participating in the market especially since vital information about the technology must be disclosed during the patent application process.

The study of patents in clean energy in China can also yield insights about Chinese technological capabilities, and whether or not they are improving in certain industries. These capabilities are directly related to firms' ability to absorb technology from abroad. For all these reasons, patents in clean energy technologies in China are examined in this section.

Measuring innovation in energy technologies is difficult. Innovation is a complex, creative process requiring human and capital investment in RD&D and diffusion that cannot be directly correlated to the production of new technologies. A range of quantitative tools for measuring energy-technology innovation exists (Gallagher, Holdren, and Sagar 2006). Scholars can count R&D programs and partnerships, R&D spending, technical publications, new lines of technology, and patents filed, granted, or cited, among others. These indicators can be classified into inputs, outputs, and outcomes.

Patents are one measure of innovative activity. Specifically, they are indicative of the output of an R&D process, but they are not without significant disadvantages, as has been thoroughly discussed in earlier papers (Basberg 1987; Griliches 1990; Pavitt 1982; Schmookler 1966). Indeed, patent analysis is susceptible to numerous problems, and three are especially relevant for this book.

The first problem is that different countries and industries have different propensities to patent. It may not be customary to patent in a certain industry or country, thereby resulting in an underestimation of innovative activity. Conversely, there may be actual firm-level or government incentives to patent, which would skew international comparisons. In China, as already discussed, the government has in fact deliberately encouraged and rewarded patent filings so we can assume that Chinese filings are somewhat inflated. One study of Chinese patent filings (overall, not limited to energy technologies) finds that university-based patent filings in China held a much higher share of total Chinese applications as compared with, for example, the share in the European Union or United States. This finding is attributed to Chinese government policies (Henning 2011). Some countries one would expect to register many patents in China appear to be underrepresented. This puzzle may stem from the fact that these countries have a lower propensity to patent, at least in China.

Second, patents granted are a measure of quantity, not quality. In China, there are three main types of patents that can be filed: invention patents, utility model patents, and design patents. Utility model and design patents are of much lower quality, and this book does not consider design patents at all. Within China, the quality of the invention patents granted could vary widely between foreign and Chinese patents as well as within both categories. Also, it is possible that SIPO's invention patent examination standards, or threshold for obviousness and novelty, may be lower than those of other countries, which would skew the results. Antoine Dechezlepretre and his colleagues (2011) interpret the relatively low rates of Chinese patenting abroad as evidence that SIPO has low standards and therefore that Chinese invention patents are of relatively "minor economic value." On the other hand, it may be that Chinese companies are not yet focused on filing abroad because they are primarily concerned with their own domestic market, which is, after all, the largest in the world. Here it is assumed that Chinese invention patents are of relatively high quality since no evidence to the contrary could be found.[14]

Third, there are many reasons why firms may decide to file a patent in China, and so a foreign firm filing a patent in China may not be a good indicator of technology transfer. Patent filers may wish to discourage imitation, augment a monopoly, crowd out potential competitors, reap royalty fees on licenses, seek remedy if their intellectual property is infringed on, obtain an injunction against a potential infringer, or export technology without fear of litigation (Suttmeier and Yao 2011; Cohen, Nelson, and Walsh 2000; Hu and Jefferson 2009). On the other hand, if a firm files a patent, it is an indication that the firm intends to participate in that market, and that there is some faith in the intellectual property regime there.

SIPO publishes Chinese patent data online in both English and Chinese. The data were collected using the Chinese-language version because the English-language version was missing records, and made country- and company-based searches more difficult. The database's advanced search function was used to disaggregate the data. Cleaner energy patents were isolated by searching for international patent classification (IPC) numbers corresponding to ten technologies of interest: coal gasification, electric energy storage, gas turbines, solar power, solar PV, advanced batteries, clean vehicles, wind, geothermal, and LEDs.[15] Domestic and foreign patent applications were distinguished by searching for foreign addresses. The publication date search function was used to break down the patent figures by year, from 1995 to 2011. Unless otherwise noted, all the patent figures in the results section refer to invention patents.

Two final limitations of the patent analysis exist. First, the search retrieves some nonclean energy technology results because the IPC codes do not perfectly isolate cleaner and more efficient energy technologies (UNEP, EPO, and ICTSD 2010). An IPC-based search yields substantially different results than one using the new European Patent Office (EPO) codes written specifically for clean energy patents. An EPO-based search using EPO's database was performed, but EPO's data from China were found to be neither complete nor up to date (EPO does not have data for 2010–2011). Since SIPO's database does not accept EPO codes, IPC codes had to be relied on.

Finally, foreign firms may be using subsidiaries to file patents from within China, and thus foreign patents are showing up as Chinese patents. Without individually checking each patent application, it cannot be determined how frequently this is happening. Assuming that it is, Chinese patent data may be somewhat inflated and foreign patent data underreported.

In sum, patents can provide insights about growing innovation in a sector or increased technology transfer, but they alone cannot be conclusive. This patent analysis is a useful complement to the case study findings, although additional case studies and surveys should be done.

Patent Trends for Different Clean Energy Technologies

For the clean and efficient energy technologies examined, few foreign invention patents were granted in China prior to 2003, and Chinese patents were negligible before 2005. Both Chinese and foreign clean energy invention patents have been increasing rapidly since 2005, however. The annual rate of Chinese invention patent growth is faster than that of foreign patent growth in all the energy technologies explored, as can be seen in figure 5.1. With respect to geothermal energy, coal gasification, and wind energy, the total number of Chinese invention patents has exceeded foreign patents since 2005. For solar PV and clean vehicles, Chinese invention patents have recently overtaken foreign invention patents. In LEDs and advanced batteries, foreign patents constituted a majority of the total number of invention patents as of 2011, but Chinese patents will soon exceed foreign patents if they sustain current growth rates. Foreign invention patents show a strong and stable lead only in gas turbines, where a persistent gap has existed since 2003.

In this section, individual results for each energy technology are provided. Table 5.2 summarizes the data presented in the technology-by-technology description below.

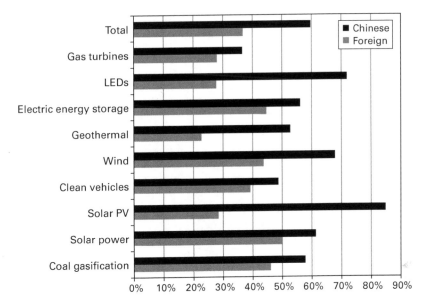

Figure 5.1
Annual average invention patent growth in China, 2006–2010

Table 5.2
Clean energy invention patents granted by China's SIPO (1995–2011)

Cleaner energy technology	Foreign percentage of total	Chinese annual growth rate since 2005	Foreign annual growth rate since 2005	Year that number of new Chinese patents surpassed number of foreign patents
Gas turbines	78%	37%	28%	Not yet
LEDs	67%	72%	28%	Likely in 2012
Wind	34%	68%	44%	2005
Clean vehicles	50%	49%	39%	2009
Coal gasification	32%	58%	46%	2003
Advanced batteries for vehicles	52%	n/a*	n/a*	n/a*
Geothermal	26%	53%	23%	2002
Solar PV	48%	85%	28%	2008

Notes: All data from SIPO; calculations by author. * The data series is too short.

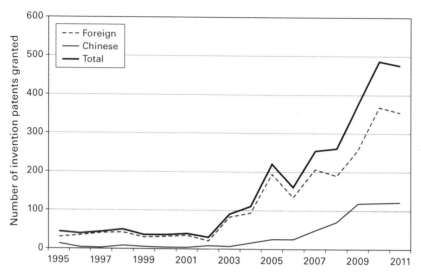

Figure 5.2
Gas turbine invention patents granted in China

Gas Turbines Foreign companies are responsible for 78 percent of the 2,753 gas turbine invention patents granted since 1995. Foreign patent numbers have been rising since 2003, but Chinese patents only began to grow significantly in 2007. Chinese patents are gradually catching up; they have increased 47 percent annually from a tiny base in 2006, while foreign patents have increased at an annual rate of 29 percent, as can be seen in figure 5.2. Gas turbines are one of only two clean energy technologies examined for which the total foreign patents still greatly outnumber Chinese patents. It is the only technology for which Chinese patents are not poised to quickly overtake foreign patents if current trends continue. US patents constitute 40 percent of the total, making gas turbines the only US-dominated area of China's clean energy technology patents. Japan contributed 13 percent, Germany 8 percent, and France 7 percent of the total.

LEDs Foreign patents constitute 67 percent of the 14,743 LED invention patents granted in China since 1995. The number of foreign patents began to rise in 2001–2002, but Chinese patents remained low until 2005. Since 2005, Chinese patents have been increasing at a rate of 43 percent annually, while foreign patents have been growing by only 11

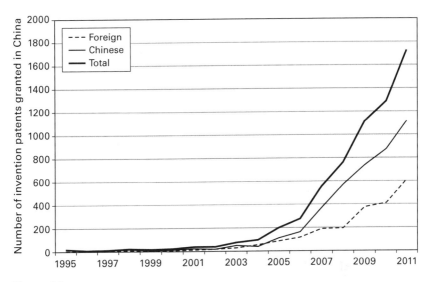

Figure 5.3
Wind power invention patents granted in China

percent per year. Japanese companies hold 29 percent of the total patents, followed by Taiwan with 15 percent and South Korea with 12 percent. LEDs and advanced batteries are the patent categories with the highest representation from China's East Asian neighbors.

Wind Energy   Foreign companies are responsible for 34 percent of China's 6,278 wind energy invention patents granted in China since 1995. While Chinese and foreign patents both began a rapid ascent in 2005, Chinese patents have since grown at a brisk rate of 52 percent per year while foreign patents have grown at 38 percent per year (see figure 5.3). SIPO approved 48 percent of the total wind energy patents in 2010–2011 alone. The United States, Germany, and Denmark are the leading holders of foreign patents with 10, 7, and 5 percent, respectively.

Clean Vehicles   Foreign patents make up 50 percent of the 9,224 clean vehicle invention patents granted in China since 1995. Foreign and Chinese patents both took off in 2005–2006 with annual growth rates since then of 24 and 51 percent, respectively. Though new Chinese patents trailed foreign ones by a small margin until 2009, they leaped

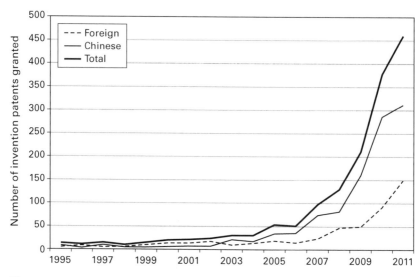

Figure 5.4
Coal gasification invention patents granted in China

up to 60 percent in 2010 and stayed at 58 percent for 2011. Japanese companies have been granted 24 percent of the total patents, and US companies have been granted 10 percent.

Coal Gasification   Foreign companies were granted 32 percent of China's 1,571 total coal gasification invention patents. Slight growth began in 2005, with Chinese patents taking off in 2007 and foreign patents following suit in 2010. Chinese patents made up 68 percent of both the 2011 patents and the total—one of the highest percentages for any technology (see figure 5.4). The United States led the foreign contributors with 12 percent of the total, followed by Germany with 7 percent, and Japan with 5 percent.

Advanced Batteries   Foreign patents constituted 52 percent of China's 683 advanced-battery invention patents in 2010. Both foreign and Chinese patents increased significantly in 2011. SIPO only started using the advanced-battery IPC numbers to classify patents in 2010. Thus, it is possible that the growth does not represent more innovation but rather SIPO transferring more patents out of other battery categories and into this one. Although the sample size is still too small for much meaningful analysis, Japan leads among foreign countries at 30 percent of the total, and South Korea follows at 11 percent.

Geothermal Energy   Foreign patents make up only 26 percent of China's 3,452 total geothermal energy invention patents. Both foreign and Chinese patents began growing from similarly low levels in 2005. Chinese patents constituted 78 percent of the 2011 patents and 74 percent of the total—the highest Chinese percentage of any clean energy technology in this study. Japan was granted the most patents of all the foreigners with 9 percent of the total.

Solar PV   Foreign companies are responsible for 48 percent of China's 17,004 total solar PV invention patents. Foreign patent growth began in 2003, two years before Chinese patents started to rise (see figure 5.5). Since then, Chinese patents have grown at a whopping 68 percent per year, for the fastest growth in any clean energy technology and far faster than the 20 percent annual foreign growth. Chinese patents overtook new foreign patents in 2008 and total foreign patents in 2010. In 2011, Chinese patents constituted 61 percent of the total. Japanese patents made up 19 percent of the total, and the United States and Taiwan both held 9 percent.

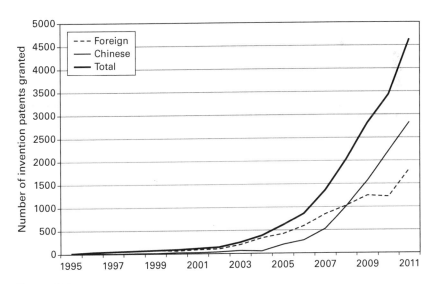

Figure 5.5
Solar PV invention patents granted in China

Foreign Patent-Holding Countries

The foreign countries holding the most invention patents in China are Japan, the United States, Taiwan, and South Korea (in that order) with a more modest contribution from Germany. Japan holds twice as many patents in the ten clean energy categories as the second-largest holder, the United States. The United States holds the dominant share of total invention patents in gas turbines. In coal gasification, the United States leads other foreign countries, although it looks small in comparison to China. Japanese patents do not exceed Chinese patents in any area, but they lead other foreign countries in advanced batteries, clean vehicles, solar PV, and LEDs. Taiwan follows Japan in solar PV and LEDs, while South Korea follows Japan in advanced batteries. In wind energy, the United States is followed by Germany, Denmark, and Japan.

Company Patents

While it was not feasible to search for the leading companies holding patents for all ten technologies, the leaders for both gas turbines and coal gasification were examined. GE alone holds 32 percent of the total gas turbine invention patents in China. Mitsubishi accounts for 8 percent and Siemens for 7 percent of the total invention gas turbine patents. In coal gasification, GE, Mitsubishi, and Siemens each hold seven patents, accounting for 24 percent of the total foreign patents. In both categories, Chinese individuals had filed most of the patents so it was not possible to determine their university or firm affiliation.

Patent Quality

Although invention patents are a better metric of innovation than utility model patents, trends in utility model patents granted in China were examined mainly because the Chinese have filed thousands of utility model patents in the clean and efficient energy realm so it is interesting to compare the patents. Utility model patents are not subjected to a substantive examination in China, and they only last ten years. Importantly, methods are not protectable because utility models are product claims.[16] Utility model patents can be granted more quickly than invention patents.

The vast majority of clean and efficient energy utility model patent holders are Chinese. Chinese companies were granted 35,940 utility model patents during the study period, while foreign firms filed only 2,993. Foreigners have been granted the most utility model patents in the LED category (20 percent of total), and the least in geothermal (2 percent), advanced batteries (3 percent), and coal gasification (1 percent).

Chinese filers had been granted 11,113 utility model patents in solar PV as of 2011, accounting for 97 percent of all utility model solar PV patents. Chinese filers had been granted 8,272 clean vehicle utility model patents as of 2011, constituting 92 percent of the total clean vehicle utility model patents. As of 2011, Chinese invention patents slightly exceeded utility model patents only in advanced batteries, LEDs, and gas turbines. Negligible quantities (less than 500) of Chinese utility model patents have been granted for advanced batteries and gas turbines, though invention patents are also low for these technologies.

Patent Discussion

The sheer number and extremely rapid rate of growth of Chinese patents in cleaner energy technologies are remarkable. Between 1995 and 2011, Chinese filers were granted 29,393 clean and efficient energy technology invention patents in China, and another 35,940 utility model patents for a grand total of 65,333. Since 2005, the average annual rate of growth of Chinese invention patents has ranged between 34 and 68 percent, depending on the technology (see figure 5.1). The quality of the Chinese invention patents is unknown, but both the scale and rate of growth indicate tremendous innovative effort, and it is well known in innovation studies that it only takes a few big hits to make a portfolio successful. Again, it must be noted that the numbers for Chinese patents are somewhat inflated due to the government incentives to patent.

As mentioned earlier, there are many rationales for registering a patent in a given country, but whether Chinese patents are being filed for offensive purposes (that is, to deter competitors) or defensive purposes (that is, to be able to protect oneself in case of an intellectual property violation), it is clear that Chinese firms and innovators expect the future Chinese clean energy market to be large and important. The Chinese are obviously developing intellectual property, and seem determined to profit from as well as protect it. Chinese firms now dominate the clean energy invention patents in China in all but one of the clean energy categories examined—a finding that was unexpected.

The widespread incidence of foreigners patenting clean and efficient energy technologies in China is somewhat surprising and intriguing as well. Overall, foreigners were granted 31,899 utility model and invention patents in China between 1995 and 2011, or 33 percent of the total (almost all were invention patents). If foreign firms felt that they had no chance of obtaining an injunction against a potential infringer, no chance of success for litigation in the Chinese court system, or no chance of remedy, they simply would not file patents in China. A number of foreign firms confirmed in interviews that in their view, it made sense to register patents in China. These firms either felt that SIPO was increasingly competent, or explained that they had actually been able to proceed with court cases, successfully litigate, and obtain compensation in China.[17] On the other hand, it also might be that these firms are just utilizing one tool of many available to them to deter competitors and discourage imitation (Pavitt 1982). Still, an improved intellectual property environment in China would create strong incentives for foreign firms to file patents given the unrivaled scale of China's likely market for clean and efficient energy in the twenty-first century.

Interestingly, foreign firms have widely different propensities to patent in the clean energy domain. There is a striking lack of patenting by German firms, which are otherwise active in the Chinese market. The United States and Japan are among the most likely to register patents in China. In comparison with an earlier Jean Olson Lanjouw and Ashoka Mody (1996) study, table 5.3 shows that the German and American percentage of clean energy patents in China has fallen over time, whereas Japanese, Korean, and Taiwanese percentages have all grown. One study of foreigners' propensity to patent in China in general finds that the surge in foreign patenting overall can be attributed to foreign firms' desire to preempt the entry of competitors. The study found that in 2004, Japan was granted the greatest number of patents and Germany the least, indicating that the country-level trends observed here are not unique to clean energy technologies (Hu 2010). By contrast, Dechezlepretre and his colleagues (2011) find Germany patented a greater share of its clean energy technologies in other countries than did the United States or Japan. If Germany files for more patents abroad than the United States or Japan, why is Germany choosing to file fewer patents in China? This is an interesting puzzle worthy of additional research.

There appears to be a relationship between the ratio of Chinese to foreign patents and Chinese technological capabilities. In technologies where foreigners hold the majority of the invention patents (say, gas turbines), China's invention capabilities are weak. Interviews with Chinese experts confirm weakness in China's gas turbine capabilities.[18] The Chinese have a high fraction of invention patents in coal gasification,

Table 5.3
Foreign invention patents as a percentage of total granted in China

	Total environmental patents (1984–1988)[a]	Alternative energy patents (1984–1988)[b]	Cleaner and more efficient energy patents, (1995–2010)[c]
Germany	5%	10%	4%
Japan	11%	6%	20%
United States	11%	16%	10%
South Korea	n/a	n/a	5%
Taiwan	n/a	n/a	8%

Notes:
a. Includes industrial air pollution, water, and alternative energy invention patents (Lanjouw and Mody 1996).
b. Includes wind, solar, waste, fuel, and heat (Lanjouw and Mody 1996).
c. Includes coal gasification, solar power, solar PV, advanced battery, clean vehicle, wind, geothermal, electric energy storage, LED, and gas turbines (author).

wind, and geothermal, but of course we cannot tell from these results if their inventive capabilities are strong. One study documented that Goldwind, China's leading wind company, was implementing a concerted strategy to register patents and defend market share, following the example of GE after its purchase of Enron Wind in the United States (Lewis 2007). Other studies have indicated that Chinese capabilities are growing strong in coal gasification (Zhao and Gallagher 2007; Chen and Xu 2010). Causality in this relationship between Chinese capabilities and the prevalence of Chinese versus foreign patents cannot be confirmed.

The 29,393 Chinese invention patents granted indicate that concrete progress is being made to achieve the Chinese government's goal of establishing a greener economy as well as its more specific targets for energy intensity, carbon intensity, and nonfossil energy use. After all, patents are a measure of the output of innovation efforts even if the quality of the invention patents is uncertain. The much larger numbers of utility model patents filed and granted by the Chinese as compared with the foreigners indicate that Chinese capabilities remain somewhat weaker than those of their competitors.

Based on the percentage of patents granted to Chinese versus foreigners, Chinese technological capabilities appear to be already strong in coal gasification, and growing rapidly in solar PV, wind turbines, advanced batteries, cleaner vehicles, geothermal, and LEDs. The only technology examined that appears to still be out of reach for the Chinese is gas turbines given the relatively low share of the total patents held by Chinese in that category.

The quality of the Chinese clean energy patents was not assessed in this analysis, and quality would be a good question to explore in some detail in future research. Of course, quality is not the same as quantity, and a "weighting" of the Chinese patents could be achieved if a quality analysis could be conducted. In general (not for energy technologies specifically), SIPO officials grant less than a third of Chinese invention patent applications, which is almost as selective as foreign patent offices are toward Chinese applications abroad. In 2010, the SIPO grant rate was 27 percent and the overseas grant rate was 24 percent (SIPO 2012). The fact that the overseas acceptance rate is this high (and has been steadily rising) seems to indicate that Chinese invention patent quality is relatively good.

Another unanswered question is the extent to which there is inflation of Chinese patents due to the financial incentives offered by different levels of the Chinese government. Local governments began to incentivize patent filings as early as 1999, but the rapid growth in clean energy technology patents in China did not occur until 2004–2005, so there is a curious lag, which might indicate that the financial incentives do not overly bias the application rate, confirming Li's analysis earlier, or the lag may indicate that the market-formation policies like China's Renewable Energy Law or the emerging eleventh five-year plan with targets for reduced energy intensity were more influential as inducements for innovation in the clean energy sector.

Overall Findings about Intellectual Property and Technology Transfer

In the clean energy sector in China, contrary to conventional wisdom, the evidence in this chapter suggests that intellectual property is not as large a barrier to the cross-border diffusion of cleaner and more efficient energy technologies. This main finding indicates that fear about this issue may therefore be distracting from more important barriers. Foreign and Chinese firms alike appear to have adopted a risk-reward mentality toward intellectual property and technology transfer.

From the point of view of foreign firms, there are clearly risks associated with transferring cleaner energy technologies to a country like China, but also substantial rewards. If they are manufacturing there, they can generally do so much more cost effectively. If they are producing for the Chinese market, there is also reward for being close to the market. To fail to participate in the largest and often fastest-growing clean energy market in the world would be to ignore a bright opportunity. Firms experienced with the global marketplace, in particular China, have developed best practices for protecting their intellectual property, such as separating production processes, compartmentalizing technical knowledge, withholding the "special sauce" that gives them their comparative advantage, organizing their partnerships with Chinese firms to maximize shared risk and reward, and having a diverse customer base. Failure to follow these practices increases the risks for foreign firms.

From the Chinese perspective, it is essential to import technology to develop and strengthen their firms. Over time, technology imports can be complemented with indigenous innovation. Many Chinese firms are not willing to put the investment in to develop strong technical capabilities due to competitive pressures. If they can license the technologies, get them into production, and then make money in the short term, that is usually the strategy they adopt. The big risks are whether the technologies they are buying will work as advertised, and whether they can absorb them into their production processes. Failure to import advanced technology is not an option for most firms that wish to remain competitive in the marketplace.

A few Chinese firms have acquired or developed strong technological capabilities over time (for instance, the professors who developed China's coal gasification technologies). These entities do not want to lose their hard-won intellectual property, and therefore, have every incentive to act responsibly and seek to partner in joint development arrangements with foreign firms. These highly capable firms are leading a relatively new trend of joint development with foreign firms or universities, or entering into global R&D consortia.

Clean energy technologies are mostly complex systems. They are hard to copy and require a lot of tacit knowledge about how they work. It requires a lot of investment in human resources and/or RD&D to acquire such knowledge. The Chinese government recognizes the importance of making these investments because it wants to shift its economy to be less resource intensive and more sustainable. Policies intended to reverse the brain drain helped to bring in expertise from around the world to guide the government's technology policy, allow China to absorb foreign technologies, and found or lead new companies. Government investments in clean energy RD&D helped to build indigenous capabilities in coal gasification and EVs in particular. Recently, major investments have begun to flow into the development of gas turbines as well. Foreign firms have also taken notice of the Chinese market for clean energy. Chinese and foreigners alike seem to have realized that without taking careful risks, there will be no rewards.

A key question is whether these findings are generalizable across technology and country. Many technologies that have military applications will clearly be more contentious than others, including nuclear energy, and these technologies are more likely to resemble the case of gas turbines. Also, Chinese firms have clearly developed strong absorptive capacities over time, which make it easier for them to bargain harder for the right technologies, and make good use of them once they are imported. Many other developing countries still lack these capabilities, much less the financial wherewithal to buy the technologies they want. Acknowledging that China is unique, the cases do provide an "existence proof" that it is indeed possible to overcome barriers as well as acquire, absorb, manufacture, and deploy cleaner energy technologies. Conversely, the fact that we do not observe a pattern of widespread infringement of cleaner energy technologies in China, the place where most foreigners would suspect infringement to be most rampant, should at least raise the question about whether governments and firms alike are overly focusing on intellectual property infringement rather than on reducing barriers and creating incentives that matter more for international technology diffusion.

6

Competing against Incumbents

The number one barrier is policy. Well, it is cost, and therefore you need to have policy to create the market.

—Shi Zhengrong, Suntech founder

Any new technology entering the marketplace must compete against existing players—the incumbents—to successfully win a share of the market. This phenomenon is true for any new technology, but cleaner technologies have additional burdens to bear because the marketplace does not naturally reward the benefits of cleaner technologies and also because access to finance for these technologies may be particularly limited due to lenders' perceptions of their risk.

Yet to effectively prevent disruptions to the global climate, improve energy security, and create more sustainable economies, the diffusion and utilization of cleaner and more efficient energy technologies must be more pervasive in the coming decades than it is today. Utilization of private markets is the only way to massively scale up the deployment of these technologies so the identification of the most important market barriers is the key analytic challenge. The higher costs of many cleaner technologies are a primary impediment to their greater diffusion. Ready access to low-cost capital can also be a constraint. Both can be partially addressed through government policy, although private firms are crucial as well.

In the case studies explored in this book, cost was identified as one of the most significant barriers to the diffusion of some of the cleaner energy technologies. Reducing or eliminating the incremental costs of cleaner technologies would make them able to compete more effectively in the market, although other challenges may persist (for instance, social acceptance or risk aversion). In many cases, government incentives around the world are proving to be sufficient to overcome the cost difference between conventional and cleaner alternatives, but the government can hardly solve these problems on its own. Indeed, the private sector must invest in RD&D, take risks, and compete in the global marketplace.

This chapter explores the nature of the challenges facing cleaner technologies in entering the energy marketplace, beginning with the theoretical understanding about disruptive technologies. Market failures are pervasive in the energy sector, and these failures are described in detail because they are the basis of the primary rationale for policy intervention. The cost and financial disadvantages for cleaner energy technologies are then analyzed in general as well as for the particular case studies.

New Entrants

Joseph Schumpeter, a pioneer in innovation studies, recognized early on that incumbent firms would almost always resist new technologies that might substitute for their products. He argued that entrepreneurs were required to bring new technologies to the market to compete against existing firms. Schumpeter (2004, 66) wrote: "It is not essential that new combinations are, as a rule, embodied, as it were in new firms which generally do not arise out of the old ones but start producing beside them. . . . [I]n general, *it is not the owner of stage coaches who builds railways.*"[1] In other words, existing firms with existing market share will always resist disruptive technologies, and rarely do they invent them from within or bring them to the market. The essential role of entrepreneurship in spurring the "creative destruction" and continuous innovation that drives economic growth has been confirmed many times since Schumpeter wrote *The Theory of Economic Development*.

Although some of the major oil companies have created renewables or low-carbon businesses (for example, BP Alternative Energy for wind and biofuels), some subsequently closed these businesses (say, BP from solar and Shell, which divested from solar, wind, and hydro but maintains a biofuels business), maintained a small stake (for instance, Chevron), or only pursued biofuels (for example, Exxon) (Webb 2009). With the exception of BP Alternative Energy along with major conglomerates like GE, Siemens, Sharp, and Panasonic, all the major cleaner energy firms are relatively new and started by entrepreneurs who were mainly devoted to cleaner energy technologies. In wind energy, these firms include Vestas, Goldwind, Suzlon, Sinovel, Gamesa, and Enercon. In solar energy, the firms include First Solar, Suntech, Sharp, Yingli, and Trina. In advanced batteries, these firms include A123 Systems (bought by Wanxiang America) and Valence Technology. All these firms are essentially competing against fossil-fuel-based firms, including major oil and natural gas companies that are both private and state owned as well as coal-mining firms and electric utilities that primarily use coal to produce electricity.

Market Failures and Distortions

Without policy intervention, the energy market sends the wrong signals to both producers and consumers. The absence of policy results in an overproduction of dirtier fuels and technologies, and an underproduction of cleaner alternatives as well as underutilization of energy-efficiency measures. Burning fossil fuels causes the release of different kinds of air pollution, including sulfur dioxide, nitrogen oxides, mercury, carbon monoxide, volatile organic compounds, particulate matter, and CO_2 (Global Energy Assessment 2012). These kinds of air pollution are damaging to human health whether from increased incidence or exacerbation of asthma and other respiratory disease, impaired neurological development, or hospitalization for heart and lung disease. Air pollution can also cause damage to the natural environment through acid deposition, smog, and eutrophication of coastal waters. The extraction, refining, and transportation of fossil fuels also can cause environmental damage, including mountain top removal, habitat destruction, oil spills, and leaching of chemicals into soils and water supplies.

All these environmental harms cost society directly and indirectly. Direct costs include health care costs including hospitalizations, doctor's visits, and medications. Even health care costs cannot be fully measured without trying to resolve impossible ethical dilemmas, such as how much to value a human life, the cost of a premature death, the value of a life free from asthma, or the psychological effects of neurological damage. Other costs are even less easily quantified, such as the effect of acid rain on a forest. The acidification may cause trees and other plants to die, which might directly harm a logging company and increase the price of lumber at hardware stores, but the damage to the forest ecosystem might also result in the loss of habitat for forest animals, or elimination of fish species from lakes or rivers. These sorts of ecological costs are difficult to quantify, even though environmental economists try to do so by estimating people's willingness to pay for the existence of pristine forests, arctic ecosystems, clean water, and fresh air.

Childhood disease, the survival of animal and plant species, and the existence of unadulterated nature are priceless in that every possible way one might quantify their value in monetary terms is highly debatable (Ackerman and Heinzerling 2004). Any effort to place prices on health, life, and nature requires making moral and ethical choices about how much to value different kinds of costs and benefits. Yet many governments do place arbitrary values on lives, health, and "ecosystem services" in order to perform cost-benefit calculations where the total estimated benefit of a proposed policy is weighed against the total estimated cost. If the benefits are estimated to exceed the costs, then the government may choose to implement the policy.

Quantification of costs and benefits is also used to determine the public or social cost of private sector activities. Once these costs are estimated, the government can better estimate the "market failure" and try to internalize it in the marketplace. The clearest example is the academic effort to determine the "social cost of carbon," which theoretically then provides a basis for imposing a carbon tax on the use of fossil fuels. The social cost of carbon is the estimated damages caused by emitting one additional ton of carbon (or CO_2) into the atmosphere. If the social cost of carbon were calculated to be $10 per ton of carbon, then the "optimal" policy would be any policy that imposes a cost of ten $10 per ton on the emitter. A carbon tax would then increase the price of the fossil fuels in accordance with the carbon content of the fuels. The more carbon in the fuel, the higher the tax, so coal would become significantly more expensive than oil, and oil more expensive than natural gas, although costs would rise in all cases because all these fuels contain carbon. Other policies could be used in lieu of a tax, including a cap-and trade program or regulation.[2]

Many economists have tried to estimate the social cost of carbon, and depending on the assumptions used, it has been estimated to range between zero and $120,000 per ton (Anthoff, Tol, and Yohe 2009). Different economists arrive at diverse estimates, however, because the calculation of the social cost requires making many assumptions and ethical decisions, such as how much to value the future versus today (the choice of discount rate). David Anthoff, Richard Tol, and Gary Yohe derive an "expected" social cost of $60 per ton of carbon, which is close to the politically negotiated price produced by the US House of Representatives in 2009. In his widely publicized review of the economics of climate change, Nicolas Stern (2006) estimated it to be $310 per tonne of carbon. Critiques of the calculations used to develop these estimates have been published (Ackerman et al. 2009; Weitzman 2009) based on the assumptions in the models and the methods used to calculate damages. Martin Weitzman (2009) argues that social cost calculation methods are inappropriate because they fail to account for the risk of catastrophic outcomes, which are inevitably impossible to predict. In his view, insurance-type schemes make more sense for the valuation of carbon emissions today. Economic damages from energy insecurity have also been estimated (for instance, Bohi and Toman 1996; Greene 2010), although they are fraught with many similar ethical dilemmas and valuation problems.

On top of the climate change damage imposed by the burning of fossil fuels, there are other social and environmental costs. Health damages from air pollution, material damages, and agricultural losses are other types of external costs. The range of health damages from air pollution and climate costs for coal-fired power plants has been estimated at approximately 7–10 cents/kWh, and 3–5 cents/kWh for natural-gas-fired plants (IPCC 2012) with large uncertainty ranges.

The point here is not to resolve these debates among economists about the precise value of the social cost of carbon or foreign oil dependence but to show that the use of fossil fuels causes economic damages that are not naturally taken into account by the marketplace, even if the estimations of these damages diverge widely. If the damages were incorporated into the price of fossil fuels, then they would certainly be more expensive and cleaner alternatives would instantly be more competitive.

Besides the pervasive externalities, a second market failure is the lack of "perfect" information. As Nobel Prize economist Joseph Stiglitz and others noted long ago, it is never possible for all participants in the marketplace to have the same access to information, and this situation reduces the efficiency of markets. In the case of both energy security and climate change, uncertainty about the future is large yet consequential. If it was known, for example, that Iran planned to cut off the Straits of Hormutz or cease oil production, then other producers would increase their supply and consumers would take steps to reduce demand so that the price of oil might not rise as much. But the impacts of climate change are inherently uncertain, the costs of mitigation and adaptation technologies in the future are also indeterminate, and the risks of energy insecurity are also unpredictable.

Firms that are considering making investments into innovation try to anticipate what kind of return to expect on these investments, but these returns are highly uncertain and also may not accrue entirely to them. As a result of this inability to fully appropriate the returns due to knowledge spillovers, firms often underinvest. An additional rationale for government investments in low-carbon innovation, therefore, is the private sector's underinvestment in this area (Nordhaus 2004; Grübler 2010; IPCC 2012).

Finally, all the market failures above are accentuated by fossil fuel subsidies, which even further distort the market. Subsidies for fossil fuels reduce their price, making the hurdle for cleaner energy technologies even more daunting, all things being equal. To make matters more complicated, some cleaner energy producers also receive subsidies in order to level the playing field and correct for the fossil fuel subsidies. While there is considerable debate about the levels of subsidies for fossil fuel producers, it is clear they are usually higher in developing countries (especially oil-producing ones) than in industrialized countries. The IEA's (2013) global estimate of fossil fuel subsidies as of 2010 was $523 billion.

The Incremental Costs of Cleaner Technologies

Unless regulations or other types of policies create incentives for the use of cleaner technologies, investors will always choose the cheapest technology in order to maximize profits. The incremental costs of many cleaner technologies can be defined as the difference in cost of the most inexpensive existing technology and fuel that can provide a specified service (say, light, cooking, mobility, or computer operations) and the cost of a cleaner alternative technology capable of providing the same service. Many government policies have already been implemented around the world that implicitly or explicitly place a value on some of the externalities. Many countries in Europe, especially, have imposed gasoline or carbon taxes, or both. The European Union established an emissions-trading regime to cap total greenhouse gas emissions and reduce them over time. India imposed a carbon tax on coal to raise revenue for energy-technology innovation. In the United States, production and investment tax credits have been extended to renewable energy technologies. In some cases, often with the help of these policies, some cleaner technologies have become market competitive. Wind energy, for instance, is cost competitive in a number of countries. Demand for gas turbines has risen substantially in the United States and elsewhere because the price of natural gas has fallen with the discovery of shale gas, and also because natural gas is much cleaner in terms of conventional air pollution as well as global climate change. In the United States, natural-gas-derived electricity is the price to beat as of 2013. In China, conventional pulverized coal power plants are the cheapest source of electricity.

Cost is a function of many factors including the price of the fuel, capital cost of the technology to convert the resource into usable energy, operation and maintenance costs, and cost of borrowing money. For the sake of simplicity, the rest of this chapter will mainly examine costs in the power sector since most of the case studies in this book are about power generation technologies (batteries can provide electricity storage in both automotive and power sector applications). The fuel costs for most renewable energy technologies are free because it obviously doesn't cost anything to obtain and use solar power from the sun, for example, but the conversion technology (for example, PV panel) can be expensive. The efficiency of the technology also affects the total cost. One of the advantages of gas turbines is that they produce electricity far more efficiently than any other fossil fuel electricity-generating technology. Coal gasification can also be a relatively efficient form of electricity generation because if coal is gasified, the synthetic gas can be used in a gas turbine and the waste heat can be used to generate steam, which can be used by a steam turbine to produce additional electricity.

RD&D efforts are often aimed at reducing the costs of these technologies. The US Department of Energy's SunShot Initiative, for example, set a goal of reducing the cost of solar PV by 75 percent to six cents/kWh through increasing solar cell efficiency, reducing production costs, strengthening the supply chain, and reducing grid integration costs, among other goals (DOE 2012d).

The capacity factor of a power generator strongly affects the cost of electricity, and one of the downsides of renewable energy is that most forms typically have relatively low capacity factors due to their variability. After all, wind doesn't blow all the time, it's only sunny during the daytime, droughts can affect biofuels, and river flows are lower at certain times during the year, thus affecting both hydropower and the cooling of thermal power plants. Some forms of renewables such as geothermal energy have capacity factors that are relatively high.

As discussed below, the cost of capital is also a major factor. If investments are perceived to be risky, the cost of capital increases. New technologies are inherently more risky when they are first deployed, which increases first-of-a-kind costs. Government policies at the national and local level can strongly affect the cost of capital. In the United States, federal loan guarantees for certain types of energy technologies have taken much of the risk away from investors and ratepayers. Feed-in tariffs, used in Europe and China, create a guaranteed return on investment, which also reduces the cost of capital.

The costs of technologies differ from one country to another, and even among regions within a single country. In the United States, the Department of Energy provides a snapshot of the current cost competitiveness of different electricity-generating technologies in its Annual Energy Outlook. Table 6.1 provides a synthesis of how costs currently compare in the US context, *not* including any policy incentives but assuming a 3 percent increase in the cost of capital for carbon-intensive technologies (because they are risky since carbon reduction legislation is likely over time). The table supplies an estimate of average costs as well as minimum and maximum costs because of regional differences as well as the effectiveness of technologies in certain contexts. Solar PV will be more efficient in sunny regions like the US Southwest, for instance. These estimated costs do not account for any policy incentives, including renewable portfolio standards, production tax credits, and loan guarantees.

Figure 6.1 converts the data in table 6.1 into a chart to make it easier to discern the incremental costs of cleaner technologies (data from EIA 2012a). Natural-gas-based power generation is obviously hard to beat even when coupled with CCS, but it should be clear that geothermal, wind, and hydropower are already less expensive than advanced coal technology (without CCS) in terms of average costs. Even utility-scale solar PV is close to competitive. While these are modeled estimates, they were benchmarked against actual costs in a 2010 study and therefore provide a reasonably up-to-date snapshot of costs in the US context.[3] A special report of the IPCC (2012) further supports this observation.

Absent government policy, if natural gas fuel prices fall further due to increased shale gas production, investors will be strongly tempted to back natural gas power plant projects in the United States. At first blush this would seem a welcome development because combined cycle power plants are highly efficient, relatively clean in terms of conventional air pollution, and produce far fewer greenhouse gases than conventional coal-fired power plants. It will be hard to justify the new construction of coal-fired power plants based on the economics, and indeed net coal-fired power generation peaked in the United States in 2007 while net generation from natural gas power plants has steadily increased since the 1990s (EIA 2012b). On the other hand, these natural gas power plants will last for decades and are unlikely to be prematurely retired. Their

Table 6.1
Estimated levelized cost of new generation resources for plants entering service in the United States in 2017 (2012$/milliwatt-hour)

Plant type	Capacity factor (%)	Fixed O&M	Variable O&M (including fuel)	Trans-mission investment	Total system cost (average)	Total system cost (minimum)	Total system cost (maximum)
Conventional coal	85	4.0	27.5	1.2	97.7	90.5	114.3
Advanced coal	85	6.6	29.1	1.2	110.9	102.5	124.0
Advanced coal with CCS	85	9.3	36.4	1.2	138.8	127.7	158.2
Conventional NGCC	87	1.9	45.8	1.2	66.1	59.5	81.0
Advanced NGCC with CCS	87	4.0	50.6	1.2	90.1	56.8	76.4
Advanced nuclear	90	11.3	11.6	1.1	111.4	107.2	118.7
Geothermal	91	11.9	9.6	1.5	98.2	84.0	112.0
Biomass	83	13.8	44.3	1.3	115.4	97.8	136.7
Wind	33	9.8	0	3.8	96.0	77.0	112.2
Solar PV (utility)	25	7.7	0	3.8	152.7	119.0	238.8
Solar thermal	20	40.1	0	6.3	242.0	176.1	386.2
Hydro	53	4	6	2.1	88.9	57.8	147.6

Source: Adapted from EIA 2012a.

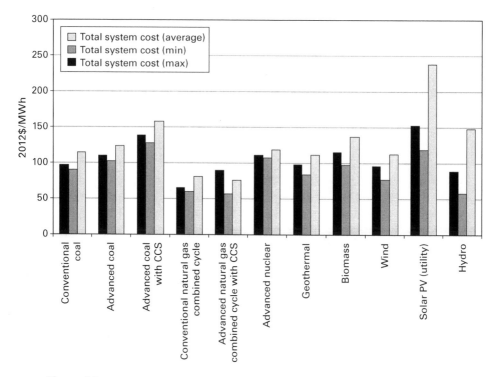

Figure 6.1
Levelized cost of new generation resources in the United States for power plants entering service in 2017. *Source:* Author using data from table 6.1.

low costs are likely to inhibit or slow down the entry of even cleaner technologies since they cannot effectively compete, they emit much greater quantities of greenhouse gases than renewable energy technologies, and the extraction of shale gas can be quite greenhouse gas intensive (Howarth, Santoro, and Ingraffea 2011). Given the economics, the resilient growth in renewable energy power generation to date can mainly be explained by policy incentives. Wind energy generation grew sixteen-fold in the United States during the decade 2001–2011 from 6,737 to 119,747 thousand MW-hour and utility-scale solar increased from 543 to 1,814 MW-hour (DOE 2012b). The role of policy was examined in detail earlier in chapter 4.

All the data provided thus far is based on costs for grid-connected power in the United States. Off-grid applications are ideal for many renewable energy sources, and they are already cost competitive in many places. Commercial and residential applications for solar PV are also able to contend on equal footing in places with good sunlight (Aanesen, Heck, and Pinner 2012). It is difficult to obtain standardized cost data across countries, with most estimates in the scholarly literature being focused on one country at a particular point in time. Interviews with US firms operating in China frequently cited a rule of thumb of a 25 percent cost reduction in China as compared with the United States for the new construction of factories or power plants. According to these interviewees, this cost reduction results from fewer permitting problems, faster construction times, a lower cost of capital, and lower labor costs for construction (but not necessarily for operation or production).

Costs over Time

The costs of cleaner energy technologies are not fixed but instead dynamically rise and fall over time. While the costs of many cleaner energy technologies have become cheaper over time, some have not. Both wind and solar have become much less expensive during the last thirty years, but other technologies have become more expensive, and during the 2000s *most* clean energy technologies became more expensive. In climate change policy debates, there are always proponents of waiting until cleaner energy technologies are cheaper. This line of argument assumes that technologies will become more affordable, yet the evidence is substantial that they do not always become cheaper. This means that we cannot assume that technologies will be cheaper with time, or that technology costs will be "bought down" through government-funded R&D, procurement, or deployment incentives. Even so, from the analysis in the previous section, it should be understood that many cleaner technologies are already competitive with conventional options. With systemic efforts to promote innovation in cleaner technologies, it is likely they would become increasingly viable as a technology option because ideally innovation will not only increase the menu of options for the future but also reduce the costs of existing technologies.[4]

One reason why we cannot predict the future costs of technologies is that we do not have a good general understanding of the factors that lead to cost reductions. The affordability of technologies is context dependent, so what works at a certain time in Japan may not easily translate to Spain, Indonesia, or the United States, and vice versa. Careful interdisciplinary research has in some cases illuminated which factors led to cost reductions in particular countries for certain technologies (see, for example, Watanabe, Wakabayashi, and Miyazawa 2000; Nemet

2006; Bolinger and Wiser 2012). In these studies, interviews with firms and government officials are complemented by analyses that examine the major factors of production and quantify their contribution to changes in price or cost. More studies of this kind are urgently needed.

More commonly, however, academic researchers calculate "learning curves" for different energy technologies. These curves estimate changes in cost as a function of cumulative production, with the rather simplistic assumption that with greater production, there is greater learning in the production process. Learning curves are usually calculated as a percent cost reduction per the doubling of cumulative output. Of course, cost reductions may occur through learning by doing (hence the term learning curve), but they can also happen for many other reasons including reductions in input prices, investments in R&D, economies of scale, government policy, and changes in market conditions (Neij 2008; Grübler et al. 2012). Most of these factors can also lead, under different circumstances, to *increased* costs too. Indeed, resource constraints and market saturation both cause learning phenomena to halt (Ferioli, Schoots, and van der Zwaan 2009).

Recent trends in costs for energy technologies are worth examining to see if the assumption of cost reduction has any basis in reality. The Global Energy Assessment provided a synthesis of learning curves as of 2011 and found negative learning (such as rising prices) for many clean energy technologies, including nuclear as well as on- and offshore wind, but reduced prices for heat pumps in Sweden, Brazilian sugarcane ethanol, and PV modules (Grübler et al. 2012). Between 2001 and 2012, retail prices for solar PV modules in the United States and Europe declined by 60 percent, from $5.50 to $2.20 per peak watt (SolarBuzz 2012). PV module prices initially fell sharply until 2003, flattened out through 2008, and then continued their sharp decline. The case study in this book explains why and how the Chinese manufacturers were able to contribute to these cost reductions. Their efforts in manufacturing, coupled with strong market-formation policies in Europe especially, explain most of the reductions in cost. The US Department of Energy estimates an average learning rate of 20 percent for solar PV since 1976 when prices fell from $60 per watt in 1976 to approximately $2 per watt in 2010 (NREL 2012). The department has set a goal of $1 per watt for its Sunshot Initiative. Gregory Nemet conducted a thorough analysis of the factors that led to the remarkable reductions in the cost of PV modules manufactured from monocrystalline and polycrystalline silicon wafers. Nemet (2006, 3218) found that "learning from experience, the theoretical mechanism used to explain learning curves, only weakly explains change in the most important factors—plant size, module efficiency, and the cost of silicon." Improvements in module efficiency had the biggest impact on module cost, followed by plant size and the cost of the major input, silicon. Substantial differences existed for installed PV systems for residential applications in different countries as of 2011, with Germany and China having average costs of approximately $3,200–3,700/kW for small systems (2–5kW) compared with approximately $5,600/kW in the United States, Italy, and Portugal (IRENA 2012a).

The story for wind turbine prices is substantially different. Between 2000 and 2008, wind turbine prices per MW *doubled* in the United States, after which time they declined 20–30 percent by 2011 due to increased competition among manufacturers and declining commodity costs (Bolinger and Wiser 2012; IRENA 2012b). Even with these rising prices, it's important to remember that the decade beginning in 2000 was the one with the greatest wind energy deployment ever in the United States, and ended with the United States having the second-largest wind energy capacity in the world (China had the largest). Wind was already competitive in the marketplace by the early 2000s due to the feed-in tariffs as well as renewable portfolio standards in the United States, China, and Europe. In 2011, wind energy accounted for 32 percent of the total capacity additions in the United States, second only to natural gas (DOE 2012a). As in the solar case, wind turbine prices vary significantly by country as a result of labor costs, local low-cost manufacturing, the degree of competition, the bargaining power of market actors, specific site factors and conditions (say, windiness), and support policies. As of 2010, China had by far the lowest average wind turbine price at $644/kW, compared with $1,234/kW in the United States, $1,526/kW in Mexico, and $2,123/kW in Austria (IRENA 2012b).

In their comprehensive study, Mark Bolinger and Ryan Wiser (2012) endeavored to uncover the drivers of the remarkable and surprising increase in costs for wind turbines for the preceding decade in the United States. They found that there were seven main drivers of changes in prices: labor costs, warranty provisions, manufacturer profit margins, turbine design and scaling up in unit size, raw materials prices, energy prices, and exchange rates. Of these factors, turbine scaling had by far the biggest effect on cost increases, followed by labor costs and then exchange rates. Warranty provisions and increases in the price of steel had small price increase effects, energy was essentially neutral, and profit margins caused a net reduction in cost impact.

Turning to gas turbines, the average estimated global investment cost of combined-cycle gas turbine plants also increased during the last decade. According to an analysis by the IEA, the costs were approximately $800/kW in 2002 and rose to $1,100/kW in 2009. These increases are attributed to high prices for materials and equipment (Seebregts 2010). The estimated cost of a large (191 MW) gas turbine alone was $220/kW as of 2008 (Pauschert 2009). No recent learning curve analyses have been published for gas turbines, nor could price data be obtained for China.

For nuclear, the experience of the French reactor program, arguably the most successful example of the deployment of nuclear power, provides a long time period to study. An analysis of changes in costs for this program found that investment costs escalated 6 percent per year between 1984 and 1990 (Grübler 2010). For the newest reactors (N4 series), the estimated construction costs were 45 percent higher per MW, which means there was a cost escalation of about a factor of 3.5 between 1974 and the post-1990 period. On the other hand, operating costs (including fuel costs) were observed to be stable throughout the time period.

As with gas turbines, no data could be obtained on the costs of coal gasification over time. In one article estimating learning curves for three methods for hydrogen production, including coal gasification, no learning curve was discerned (Schoots et al. 2008). In interviews, Chinese experts asserted that the new Chinese gasifiers were approximately 20 percent less expensive than foreign gasifiers in the Chinese market, mainly because the Chinese charge lower license fees. In fact, the Chinese gasifiers currently have higher capital costs than their European or US equivalents, but these are expected to decline as manufacturing output increases.[5]

So despite the widespread perception that costs usually decline over time, we see that in fact, costs for quite a number of cleaner energy technologies have been rising. Given the rather robust literature on learning curves, which has led many to believe that costs will decline, how do we explain these recent trends? Many answers and hypotheses are beginning to emerge (see the studies cited above; for a thorough review, Neij 2008; Schoots et al. 2008; Kobos, Erickson, and Drennen 2006). The most significant factor appears to be the increased risks for new technologies, especially as they get larger and more complex. Health, safety, and environmental regulations also cause plants to become more intricate and expensive. Two examples of the effects of complexity would be large offshore wind and nuclear power. Also, oligopolistic market structure is defined as having relatively little competition among manufacturers. A lack of competition can reduce the market pressure to cut costs. Uncertainties about government policy and approvals for permits can be large factors as well. Access to finance (related to risk) is a constraint, as are skilled labor shortages. Input costs can rise and fall, strongly affecting final product prices. The efficiency of the technology can also affect its overall affordability.

Like a Snowball Rolling Down a Hill: Access to Finance

There is no evidence in the case studies, scholarly literature, or popular press that access to finance is a barrier for Chinese firms.[6] Readily available, low-cost capital does represent a major challenge for firms in many other countries, however. The IPCC (2012, 882) Special Report on Renewable Energy observed that "financing is critical in every stage of technology deployment. Yet there are also many barriers that affect the availability of financing." These barriers include their lack of economic competitiveness (causing investors to balk), a lack of sufficient information, the higher investment costs of most low-carbon energy sources (even though renewable energies have lower operating costs), the lack of experience with new technologies, and the limited track records of many renewable energy developers. One way of interpreting these findings is that the ease of obtaining financing in China coupled with the typically relatively low cost of capital there are major explanations for why Chinese firms have made significant strides in developing their cleaner energy industries and deploying them around the world. As one US executive commented, "Very attractive financing has become a commercial advantage."[7] If governments in other countries could also expand access to low-cost financing for cleaner energy technologies, an acceleration of the global diffusion of cleaner energy technologies would likely occur.

The Chinese government has made access to low-cost capital a central component of its industrial policy for clean energy. While the central government has some control over lending, limited evidence from the cases indicates that most of the capital for China's clean energy firms comes from provincial and local governments.[8] These governments provide low-interest loans for land, electricity, and equipment.[9] Once the government supplies these loans to firms, other sources of financing materialize. One Chinese battery company official disclosed that he had more potential investors than he could utilize, for example.[10] In short, success in raising capital from the government breeds success in raising even more. This process is akin to a snowball rolling down a hill. Under the right conditions, the ball will pick up snow as it starts to roll, and accumulate more and more as it goes. Implicit but unspoken in the interviews is that a government loan is more than just that—it is an indication of government support that enables firms to accrue far more capital than they would be able to otherwise.

Access to finance is a well-known barrier to the diffusion of clean energy technologies (IPCC 2000). Clean energy projects often appear riskier to investors because new firms can lack strong credit records, the technologies themselves are less well proven or are unfamiliar commercially, and the technologies frequently rely on policies that could be changed (Goldman, McKenna, and Murphy 2005; Hamilton 2009). The barriers can be particularly acute for energy-efficiency technologies. Interest rates are not usually the same for energy producers and consumers, and the difference has been identified as the "interest rate gap" (Brown 2001). Financing constraints for rural energy users at the "base of the pyramid" are even more severe since capital markets are usually not available to the very poor. Governments and intergovernmental organizations such as the Global Environment Facility usually must step in to fill the gap, but these financial resources are nowhere close to meeting the full demand, and they are usually not sustainable financing models in the longer term (see, for example, Gallagher, Siegel, and Strong 2011).

Insights from the Case Studies

Chinese firms trying to compete in the marketplace for cleaner energy technologies must consider the cost at home and abroad. China is the largest single energy consumer in the world, and so it has a large domestic market for energy technologies. Market failures and distortions like fossil fuel subsidies have already been discussed in general. As a Chinese firm looks to foreign markets, it must evaluate these factors on a country-by-country or even locality-by-locality basis, precisely because policy interventions may have helped or hindered the formation of a market for cleaner and more efficient energy technologies. Likewise, foreign and local firms assessing the Chinese marketplace must evaluate the market in China. Because China's energy system relies most heavily on coal, conventional pulverized coal-fired power is the price that must be beat in the electricity sector.

While the government has strongly emphasized energy efficiency, it has not yet imposed any taxes on energy or carbon at the national level, although they are being experimented with at the local level. Instead, the government has developed a renewable portfolio standard, established feed-in tariffs for both wind and solar, announced targets for nonfossil energy and natural gas consumption in its twelfth five-year plan, and provided purchase incentives for EVs. It is also supporting through its RD&D programs one large demonstration program for IGCC coal-fired power, which will capture and store carbon in its second phase. At times the Chinese government has strongly subsidized fossil fuels, but it has been steadily reducing those subsidies and letting market prices prevail for the last fifteen years. The price of electricity is still controlled, yet the underlying prices for coal, gas, and oil are usually allowed to mirror the market.

In chapter 3, the higher costs of cleaner energy technologies were identified as one of the few barriers that firms of all nationalities agreed on. The cost barrier seems to be most acute for advanced-battery technologies, serious for coal gasification for electric power, and less of a problem for solar PV due to the market-formation policies around the world that have helped to make solar PV modules more competitive. In the Chinese market, the cost barrier for gas turbines is significant, although this is not true in other markets. Everyone who was interviewed stated that due to relatively high costs, policy intervention was required. The issue of access to financing sharply divided Chinese and foreign firms. In general, Chinese firms did not feel that access to finance was a major constraint for them. For US firms in particular, access to finance was identified as a central challenge.

In advanced batteries for automobiles, both Chinese and foreign firms agreed that cost was the number one barrier. As one Chinese manufacturer said, "In China, the most important barrier is price."[11] The Chinese government and its firms have struggled with how to acquire as well as deploy cleaner, more efficient vehicles for decades. From the Chinese standpoint, the cost of foreign HEVs is prohibitively high, and Toyota has refused to license its HEV technologies to Chinese producers.[12] For these reasons and others described in the case study, the Chinese have made a strategic industrial decision to focus mainly on EVs instead of HEVs or other alternative-fueled vehicles. The market niche that they see as having the most initial potential is rural, low-speed EVs that would not require advanced batteries. Because of these attributes, their costs would be much lower than a car requiring an advanced battery and faster charging times. The government has created incentives for Chinese manufacturers to sell EVs to greatly reduce the purchase price for consumers. According to industry experts, the cost for a li-ion battery in China without government subsidies is 4–5 RMB/Wh, compared with only 0.6 RMB/Wh for lead acid batteries. This would translate into a total cost for an advanced-battery set for a car of 100,000–120,000 RMB, significantly more than all the other parts and components, which are estimated at less than 70,000 RMB. In other words, the total cost of an advanced-battery vehicle in China is between 220,000 and 230,000 RMB, of which approximately half comes from the cost of the battery itself. With a lead acid battery pack, the final price would be seven to nine times lower. The Chinese automakers expect that with economies of scale, the total cost of advanced-battery vehicles would be reduced by about one-third. The government subsidy is 60,000 RMB for an electric passenger car and 600,000 for an electric bus.[13] For foreign firms, the high costs are what drove them to invest and manufacture in China in the first place because China is a less expensive place for manufacturing. The other attraction of China for the foreign firms is that financing is more readily available there. One prominent US battery firm, A123, entered into a memorandum of understanding with Chinese firm Wanxiang in August 2012 precisely because Wanxiang was prepared to invest $450 million into A123 (Ramsey 2012). In the end, Wanxiang was able to acquire A123 out of bankruptcy.

Costs remain high for coal gasification for electricity production. GreenGen will be China's first IGCC plant, and the anticipated price of electricity from it will be double the average price for a normal coal plant in China. One way that GreenGen was able to save money was to use the Chinese TPRI gasifier, which was estimated to be 20 percent cheaper than its foreign equivalent, mainly because GreenGen did not have to pay a license fee since TPRI is owned by the same firms investing in GreenGen, most notably Huaneng Group (the lead investor), but also because they could use more domestic materials.[14] A central point about gasification is that IGCC technology is complex and foreign, and Chinese firms alike have found it technically difficult to reduce costs.[15]

In the case of solar PV, the cost of acquiring foreign technology was not prohibitive for Chinese firms. These firms' remarkable ability to manufacture at increasingly lower cost has been a source of considerable controversy with some foreign firms accusing the Chinese of dumping PV modules in foreign markets or at least being illegally subsidized—

accusations that the Chinese have vigorously denied. According to interviews, there are four main sources of cost reductions. The first is the relative speed of the Chinese manufacturers along with their related ability to ramp production up and down in response to market demand. The second source is efficiency, in the sense of both pushing manufacturing efficiencies to the limit and improving the technical efficiency of the solar cells to reduce the use of silicon .[16] Shi Zhengrong, CEO of Suntech, holds to the personal motto of "thoroughness," and he carried it into his factories to identify every possible way to reduce costs. In fact, in a speech at MIT, he confessed that when he developed the business plan for Suntech, he had no idea how to estimate costs so he just imposed a 30 percent discount on everything, which, he asserted, turned out to be accurate. Reductions in the cost of polysilicon and increased wafer efficiency also caused price reductions for firms of all nationalities. All the Chinese solar PV manufacturing plants visited during the course of this research were highly automated so low-cost labor cannot be a major source of cost reductions except during the construction phase of a factory. Finally, vertical integration allows firms to eliminate "margin stacking."

In the case of gas turbines, there were highly divergent views about their costs. The Chinese perceived the costs of advanced heavy-duty gas turbine technologies to be high, both in terms of final costs and licenses. Less advanced (and less efficient) turbines are more economical and within reach for the Chinese. Coupling the high costs of the technology with the relatively high prices that operators pay for natural gas in China (due to the fact that it is mostly imported), it is easy to see why natural gas power generation has remained extremely limited in China. To the foreign producers of gas turbine technologies, the costs are reasonable given the R&D that has been put into the latest turbines. To the Chinese, because the costs are so high, they have no choice but to earnestly innovate to be able to produce at a lower cost. As one Chinese scientist commented, "China cannot suffer these extremely high costs so China must develop its own technology."[17]

Summary

The cost of cleaner energy technologies is the most important barrier in many cases. Although some energy technologies have become cheaper as a function of cumulative production, during 2000–2010 costs started to rise in a worrying manner for quite a number of cleaner energy technologies. Access to finance, particularly low-cost capital, is not found to be a barrier in the Chinese cases, but evidently is one of the most important reasons why Chinese firms have been able to acquire, further develop, and then sell cleaner energy technologies around the world. Reasonable cost and access to finance is therefore a critical catalyst for the greater diffusion of clean and efficient energy technologies worldwide.

7

The Global Diffusion of Cleaner Energy Technologies

This chapter synthesizes the theory and evidence regarding the global diffusion of cleaner and more efficient energy technologies. The first conclusion is that during the 2000s, the clean energy sector experienced a pronounced globalization in both the development and deployment of cleaner energy technologies. The main drivers of this globalization, which continues today, are the internationalization of postsecondary education, ease and increased normalcy of migration, establishment of new national- and subnational-level policies to address pollution, energy security, and global climate change, thereby creating global clean energy markets, and creation of international institutions that facilitate cross-border investments and trade.

After exploring this process of globalization, the chapter examines the aspects of technology transfer theory that no longer seem to fit well with the emerging evidence, including the evidence within this book, about the diffusion of cleaner energy technologies. The main findings are that international technology transfer is no longer a unidirectional north-to-south process; innovation in cleaner energy technologies is no longer primarily a national process but rather one that is highly globalized; the global diffusion of cleaner energy technologies requires extensive national or subnational market formation by governments; most technology transfer occurs through private markets, not public institutions; and many of the often-assumed barriers to the diffusion of cleaner energy technologies, including higher costs, access to technology, and infringement of intellectual property, among others, are not insurmountable. The lack of market-formation policies, inadequate access to low-cost capital, higher costs of some cleaner technologies, and unwise business practices all present formidable challenges, but not absolute barriers to the global diffusion of cleaner energy technologies. A new integrated theory of the global diffusion of cleaner energy technologies is then offered to help explain why there is no "great wall" in the global diffusion process.

Given that this book presents evidence from China, it must be restated here that in many ways, the Chinese experience is unique and cannot be easily replicated. Five features stand out in particular. First, the sheer size of China's economy makes the localization of many cleaner energy technologies in China sensible for Chinese and foreign firms alike. Everyone wants to be close to such a large market. Second, the remarkable savings rate along with the related ability as well as willingness of the central and local governments in China to finance large investments in cleaner technologies is probably unrivaled. The Chinese have a longer-term perspective and greater willingness to direct investments into clean energy than do most other governments. Third, China's political system and traditions are unusual. One big advantage in this regard is the strong historical emphasis on long-term, regular planning (although it is frequently asserted that China's centralized government structure allows the government to impose its will, there is not much proof that this is true). Fourth, China has a highly productive, relatively low-wage, and large labor force. Fifth, China's infrastructure is quite good compared with that of developing countries, making it relatively easy for its firms to import and export technology.

Yet this book provides concrete, empirical evidence that countries, developing or industrialized, can acquire, modify, develop, manufacture, and export cleaner energy technologies. Most countries will not attempt to do this for so many technologies at once, but the Chinese experience does offer a model for how a country might exploit a comparative advantage to localize an industry or embrace a foreign market opportunity. Other countries have shown they can likewise acquire cleaner technology and render productive industries—the wind industry in India is an excellent example, as is the sugarcane ethanol industry in Brazil. At a minimum, this book shows how countries can certainly achieve greater *deployment* of cleaner energy technologies within their borders, if not localization. From the point of view of preventing global climate change, it is the global deployment of cleaner and more efficient energy technologies that is required. Some countries may wish to capitalize on the trend toward the green economy or cleaner energy systems, in which case they may endeavor to become producers of these cleaner technologies. China has managed to do both.

The Globalization of Cleaner Energy Technologies

Cleaner energy technologies are much more globalized today than they were only two decades ago. To be precise, cleaner energy technologies are being developed in numerous countries using a global knowledge base, and then diffused through global markets shaped by national and subnational government policies. The decade beginning in 2000 appears to be when the transformation from national to global occurred. A number of drivers of this change are identified below, but new market-formation policies and the emergence of China itself in the clean energy sector explain most of the shift.

As discussed in chapter 5, during the late 1990s and early 2000s a large number of market-formation policies were established in support of cleaner and more efficient energy technologies. The rationales for these policies varied widely, but all caused new markets to spring up, which were noticed by observant entrepreneurs. These policies were varied in accordance with different national circumstances, and some were experimental in nature. Over time, lessons have been learned about effective policy design, and more recent policies reflect the knowledge gained from past experience (see, for example, Cozzi 2012). Scandinavian countries were the early pioneers of most clean energy policies, imposing carbon taxes and feed-in tariffs for renewables in the late 1980s and 1990s. Most large industrialized countries did not enact major market-formation policies until the late 1990s, perhaps inspired by the establishment of the Kyoto Protocol to the UNFCCC in 1997. China's Renewable Energy Law, the EU Emissions Trading System, Germany and Spain's feed-in tariffs for renewable energy, Brazil's National Climate Change Policy, India's carbon tax, Japan's Top Runner program, and the proliferation of renewable portfolio standards at the state level in the United States all contributed to the establishment of a global market for cleaner, more efficient energy technologies. During this period, China became one of the largest—and in some years, *the* largest—investor in renewable energy technologies. According to the Pew Charitable Trusts (2013), in 2012, China attracted $65.1 billion in investment, 20 percent more than in 2011 and an unmatched 30 percent of the G-20 total.

One indicator of the extent and rate of globalization is the volume of world trade. As shown in chapter 4, the volume of global imports and exports grew dramatically during the 2000s, especially for solar and wind technologies. Exports of cleaner energy technologies from China and other emerging economies have grown substantially since the late 1990s. The total volume of trade in clean energy expanded rapidly—more than twice as fast as manufactured goods overall. For four clean energy technologies, the volume of exports and imports grew 259 percent between 2000 and 2010, compared with 118 percent for the exports and imports of the total manufactured goods globally.[1] China accounts for a huge proportion of the growth in global trade. Chinese exports of solar PV were only $93 million in 1997, accounting for just 3 percent of the global solar exports, but in 2011 Chinese exports had soared to $27 billion, accounting for 45 percent of the global exports. China is far from being the only developing country contributing to the global marketplace in cleaner energy technologies. Together with Indonesia, Malaysia, Mexico, Poland, and Hungary, for instance, China accounts for 20 percent of the global LED exports. The growth in volume of trade for both solar and wind lags by a few years the enactment of new policies around the world to stimulate markets in renewable energy, indicating that after new policies are established, producers initiate or ramp up their production for export.

The data presented in chapter 5 on clean energy patent growth in China is yet another source of information for when the globalization of these technologies deepened. In every technology examined (gas turbines, LEDs, wind power, clean vehicles, coal gasification, geothermal, and solar PV), a sharp uptick in the growth of invention patents granted occurred between 2002 and 2006, with most turning points happening in 2005.[2] The annual growth in foreign patents granted in China from 2006 to 2011 averaged 21 percent, and 52 percent for Chinese patents granted. Again, we see a striking coincidence in growth in foreign patents granted in China with the market-formation policies enacted in the early 2000s. Overall, foreigners were granted 31,899 utility model and invention patents in China between 1995 and 2011, or 33 percent of the total.

The Drivers of the Global Diffusion of Cleaner Energy Technologies

The preceding section described a relatively new phenomenon of the globalization of both the development and deployment of cleaner energy technologies. This section presents seven key drivers of this globalization process: the internationalization of university education; increased international collaboration among firms, universities, and research institutes; the ease and increased normalcy of international migration; the emergence of a global energy technology innovation system; the proliferation of national market-formation policies; trade liberalization and new international institutions; and China's willingness to finance major investments into cleaner energy technologies.

Internationalization of University Education

International education is surely a major cause of the globalization of cleaner energy technologies. According to the UN Educational, Scientific, and Cultural Organization (2011), between 1998 and 2010, the number of internationally mobile students nearly doubled. China's share of the total grew from just 7 percent in 1998 to 17 percent in 2010. No other country comes close to China's volume. The Chinese solar industry case shows how important educated Chinese returning home from overseas are proving to be to the growth of the Chinese clean energy economy.

International education not only improves the technological capabilities of individuals and their countries but it also provides a global perspective and fosters the development of international learning networks. Innovation theory emphasizes the benefits of diversity (Lester and Piore 2004) as well as the utility of learning by interacting (Lundvall 1988) through international learning networks (Ernst and Kim 2002). The more international learning and networking, the more rapid the progress of innovation is likely to be (Grübler, Nakicenovic, and Victor 1999). International learning networks and collaboration are important in both the solar and coal gasification cases, and were found to be crucial for the development of the Chinese and Indian wind industries as well (Lewis 2007). The lack of international collaboration in the gas turbine industry appears to be a contributing factor in its relative weakness.

International Collaboration

International collaboration among firms, universities, and research institutes greatly facilitates the cross-border transfer of hardware as well as clean energy know-how. This firm-level cooperation can be effectively catalyzed and fostered by government agreements, but government involvement does not appear to be necessary. International collaboration takes many forms, and can include strategic alliances, foreign direct investment, formal joint ventures, contracts, trade, joint development of technology, and licensing. Chinese and foreign firms have experimented with all these forms of collaboration. To provide a few examples from the cases, US firm FutureFuels has licensed the Chinese Thermal Power Research Institute coal gasification system for use in the United States. Duke Energy formed a memorandum of understanding with Huaneng in

2009 to do joint research and development on IGCC, CCS, efficiency, and renewable energy technologies. Suntech does joint R&D on solar energy technologies with the University of New South Wales in Australia, China's JA Solar has a joint development agreement with US firm Innovalight, and China's Trina Solar and DuPont have agreed to collaborate on advancing the efficiency as well as lifetime of solar cells and modules. Yingli Solar has a joint development agreement with the Energy Research Center of the Netherlands.

Other examples of international clean energy technology cooperation can be found in the wider literature. India's famous wind turbine manufacturer, Suzlon, has bought several foreign firms, including a rotor blade manufacturing company in the Netherlands, and established its world headquarters not in Dehli or Bangalore but rather in Aarhus, Denmark, to take advantage of the wind energy expertise there (Lewis 2007).

The Ease and Increased Normalcy of Migration

The ability of people to migrate around the world greatly increases the flow of knowledge and broadening of perspective. This phenomenon is critical for international education, as discussed above, but also for the acquisition of business experience in markets of interest. Firms that have succeeded in the Chinese market have deployed people who have spent many years acquiring experience there. The former president of BP China, Gary Dirks, lived in China for fourteen years and was eventually recognized by the *People's Daily* as one of the ten most influential multinational leaders in the previous thirty years of development. Likewise, many individuals in Chinese firms have spent time abroad acquiring experience. Indeed, one of the most ardent advocates of cleaner vehicle technologies, the current minister of science and technology Dr. Wan Gang, completed his doctorate in engineering in Germany and stayed on to work at Audi for nearly a decade. International migration can strengthen human capital and contribute to "brain circulation" rather than brain drain (Trachtman 2009, 60).

Increasing Globalization of Clean Energy RD&D

Energy RD&D is no longer the purview of industrialized countries alone; the entry of developing countries is both adding to and diversifying the global effort. The Global Energy Assessment of 2012 provided the first-ever global estimate of total energy RD&D, and it arrived at a total of $50 billion per year at a minimum (Grübler et al. 2012). Of that total, at least $14 billion comes from government investments in the major emerging economies according to an estimate developed by Rudd Kempener and his colleagues (2010), and public investments in China alone account for $12 billion. The emerging economies thus account for almost one-third of the global energy RD&D investments. On the whole, however, it appears that most of the global energy RD&D effort is on the supply side, split roughly equally between fossil fuels ($12 billion), nuclear ($10 billion), and renewables ($12 billion). In contrast, only $8 billion is devoted to RD&D on end use and efficiency (Grübler et al. 2012). From these figures, then, the global investment in cleaner energy technology RD&D is at least $20–42 billion, depending on whether or not nuclear and cleaner fossil fuel technologies are included in the total.

In the private sector, multinational corporations do RD&D and source technology development from around the world. Developing country firms now acquire industrialized country firms as a strategy for technology acquisition, and vice versa. Consultants from around the world are hired to do debugging work and find solutions to technical problems.

Aggregated National Policies Equals Global Markets

New national and subnational policies to address pollution, energy security, and global climate change created global markets for cleaner energy technologies beginning around the turn of the twenty-first century. Firms can take advantage of these opportunities if they have the global perspective to see them. Policies in Germany, Spain, the United States, and China were particularly significant for the diffusion of coal gasification, solar, and advanced batteries.

Interviewee after interviewee ranked policy as the most critical incentive for the deployment of clean energy technologies. The chair of Shell China, H. K. Lim, emphasized the importance of policy to create certainty in the clean energy market because it allows firms to calculate a return on investment. He said that as a businessperson, if he knows the return on investment is attractive, he "can't wait" to get into that business. Ed Lowe of GE commented, "Government policy is extremely important to drive long-term sustainable development," and stressed that support for both R&D and domestic manufacturing were essential to proving that a given technology is viable. Hans-Peter Böhm of Siemens observed, "Without government regulation, you won't have a market for clean energy." Li Wenhua, formerly of GE China said, "IGCC relies heavily on policy. If there is no carbon dioxide policy, there is no incentive." Tim Richards of GE remarked, "If leaders don't put an economic value on emissions and security, but want to see higher levels of renewables and lower emissions, you need smart policy that creates clear incentives. The same is true for coal gasification, and especially CCS. By contrast, because of current gas prices, natural gas-fired power projects are economically viable within the current policy framework." Shi Zhengrong, CEO of Suntech, asserted that "the number one barrier is policy. Well, it is cost, and therefore you need to have policy to create the market." Dick Wilder of Microsoft was more specific: "The policy environment is important—principally the stability and the predictability. Especially when you are talking about large installations when the payback will take place over a long period of time, you want to be sure you have a regulatory environment that will be predictable while you are expecting to recoup your investment." Most of these quotations refer to market-formation policies, and these policies need to be predictable, stable, and aligned (Grübler et al. 2012).

As discussed in chapter 5, several other types of policies can facilitate the international diffusion of cleaner energy technologies as well. Industrial or manufacturing policies allow firms to demonstrate and scale up new technologies at home. Export promotion policies can help overcome barriers to trade, such as inadequate financing. Policies that encourage (or conversely reduce barriers to) trade and international migration also facilitate the cross-border transfer of clean energy technologies. Finally, policies that create a robust regime for the protection of international property can create a more welcoming environment for new technology. Michael Rock and David Angel (2005) note that due to China's deep integration into the world economy through global production networks and export-oriented industrial production, China is especially "attuned to the sensitivities of global markets" including the emergence of significant end-market regulation, which lends support to the conclusion that economic globalization is spurring the Chinese to participate in a global clean energy market.

Trade Liberalization and New International Institutions

The international institutions that facilitate cross-border investments and trade are much more sophisticated than they were two decades ago. The WTO facilitated the expansion of trade and accelerated China's participation in the global economy, which in turn forced Chinese firms to become much more competitive. China's entry into the WTO also gave Chinese firms greater access to global markets. It is almost unthinkable that China's solar PV industry would have been able to acquire all the manufacturing technology it needed to build an industry in only a couple of years and then capture the global market without China's membership in the WTO. Maybe China hasn't played by all the rules, but it has demonstrated what determination and a high level of ambition can achieve—with plenty of money.

Not a single interviewee for this book even mentioned the WTO or TRIPS as facilitating or hindering cross-border technology transfer, so there is no evidence from the case studies that these institutions play a strong role either way. On the other hand, it is likely that people now take these institutions for granted since the WTO and TRIPS came into existence in 1995. Although China did not enter the WTO until 2001, a decade has already passed, and China's membership is the new norm. It is often noted in the literature that China made numerous changes to its intellectual property laws to conform to the TRIPS agreement, amending its patent, copyright, and trademark laws in 2001 (and subsequently amending these laws again).

Due to the wide disparities among national laws with regard to enforcement, TRIPS only provides minimum standards for the enforcement of intellectual property rights (UNCTAD-ICTSD 2005), and this could be another explanation as to why the WTO and TRIPS do not figure strongly in this analysis about the movement of cleaner energy technologies to and from China. This outcome is somewhat ironic because one of the rationales for TRIPS was to improve the effectiveness of intellectual property rights obligations in international agreements. The global norms around intellectual property rights—codified in TRIPS—have created strong incentives for firms to transfer technology. Even in China, a pretty scary place for most foreign firms, foreigners filed and were granted 31,899 clean and efficient energy patents between 1995 and 2011. But while the patent regime is clearly growing and deepening in China, undoubtedly less progress has been made on enforcing intellectual property laws.

Even with the existence of the WTO and its TRIPS agreement, it is important to recognize that no explicit formal international agreement exists to govern or even guide international technology transfer. A serious effort to establish such an agreement was made during the 1970s when nations endeavored to negotiate an International Code of Conduct on Transfer of Technology under the aegis of the UN Conference on Trade and Development. These negotiations continued for ten years, but no agreement was reached. A number of issues contributed to the impasse: objectives and principles, restrictive business practices (antitrust, antimonopoly, and competition laws), performance requirements on foreign investments and technology transfers, dispute settlements, intellectual property protections and compulsory licensing, and whether the code would be voluntary or mandatory. While the code of conduct did not ultimately prevail, some of the issues were taken up and resolved in the TRIPs and Trade Related Investment Measures negotiations (Patel, Roffe, and Yusuf 2001).

Chinese Willingness to Finance the Transition

The cases show that the Chinese government at both the central and local levels has made an unprecedented commitment to finance a transition to cleaner, more efficient energy technologies. The provision of low-cost capital has not only contributed to the domestic diffusion of cleaner technologies within China but also accelerated the global diffusion of cleaner technologies. The case of the Chinese solar PV industry provides the clearest example because the Chinese firms were not initially targeting the Chinese market but rather planned to export to the German, Spanish, and US markets. In order to build their factories and scale up manufacturing, the Chinese firms required substantial capital, which was readily available to them in China. Bloomberg New Energy Finance and the Pew Charitable Trusts (2012) have tracked clean energy investments for a number of years now, and China has become one of the leading investors, along with the United States, Germany, and Italy.

The Contemporary Understanding of the Global Diffusion of Cleaner Energy Technologies

Some aspects of technology transfer theory do not seem to fit with the emerging evidence about the diffusion of cleaner energy technologies. International technology transfer is no longer a unidirectional north-to-south process (Brewer 2008). It should also be clear by now that innovation in cleaner energy technologies is no longer primarily a national process. Technology is not developed in a single country and then physically transferred to another. Rather, technologies are acquired from all corners of the globe, using a diverse set of mechanisms. We know that technology is cumulative and iterative, and the incremental development of cleaner energy technologies takes advantage of a global knowledge base. Global learning networks now exist, and major emerging economy firms are actively cultivating these networks.

The cases in this book show that cleaner energy technologies can be acquired from all over the world, further developed and modified, and then final products diffused through global market forces that are strongly shaped by government policies in many countries. Entrepreneurs, multinational and domestic firms, and research institutions including universities are highly global in their orientation. It is not that they do not respect national borders but instead that they have a global perspective. Chinese firms have utilized literally every mechanism for the cross-border transfer of technology, and so have their foreign partners.

In the subsequent sections the main theoretical concepts that no longer seem valid are reviewed before turning to a new integrated theory of the global diffusion of cleaner energy technologies.

Bigger Is Better

Although the concepts in Ernst F. Schumacher's (1973) seminal book *Small Is Beautiful* are attractive, in the cleaner energy domain, it is not at all clear that smaller, individualized technologies are better for developing countries. The notion that technologies must be appropriate for individual-scale needs and conditions connotes a kind of approach to technology transfer that is, first, north to south in concept as well as almost *artisanal*; the interpretation is that extensive adaption and modification of technology to suit each application is required. There is little evidence of the need for such a small and piecemeal approach in these cases. Some technologies did need to be adapted for the Chinese context, most strikingly for coal gasification technologies. But once they were adapted to Chinese coals, they could be scaled up and used everywhere in China. Interestingly, the new Chinese gasifiers appear to be more flexible for use outside China too, as evidenced by the US license of the TPRI gasifier. Other than for coal gasification (where the coal characteristics were at issue), size and lack of "appropriateness" was not a serious barrier for the other technologies.

In the context of climate change, it could be argued that small is imprudent and bigger is better. Cleaner energy technologies cannot be globally deployed with government-sponsored, small-scale, individualized technology transfer initiatives, project by project. If this approach is taken, it is unlikely that emissions will be reduced at a scale and within a time frame that avoids serious temperature and other climatic changes. The global scaling up of markets allows for a more rapid diffusion of technologies, and relatedly, a reduction in the costs of these cleaner technologies, which in turn makes them ever more accessible to the poorer countries. To provide just one example of this potential, consider the significant declines in solar PV module prices in the past five years that resulted from market-formation policies in Europe and the United States along with Chinese manufacturers entering the market. These cost reductions make solar PV more attractive for all consumers and within reach for many developing countries. Consequently, the Chinese government launched a new program in 2012 to invest $100 million in solar projects in forty African nations using competitive bidding and Chinese-made panels (*China Daily* 2012a). Before this program began, Chinese manufacturers were already exporting $130 million worth of solar panels to the African continent annually as of 2011 (Comtrade 2012).

An increasing market scale appears, then, to be critical to the global diffusion of cleaner energy technologies. If demand for these technologies exists anywhere, it can be met through suppliers in many different countries. Demand for cleaner technologies is mainly created through market-formation policies at the national and subnational levels, as discussed below and in chapter 4. With more commercial experimentation, establishment of an industrial base, and increasing standardization for mass production to meet such demand, costs may fall. Economies of scale are not the only sources of cost reductions, however (Argote and Epple 1990; Nemet 2006; Grübler 1998).

It is important to distinguish between the smallness and bigness of *market* versus *unit size*. It is not necessarily desirable to scale up unit size, although there can be advantages in doing so (Wilson 2012). Diseconomies of scale can occur due to increasing technological complexity and other factors (in the case of French nuclear power, for example, see Grübler 2010). Production of many small units can be helpful because the smaller scale allows for extensive experimentation and learning, and provides a relatively flexible way to meet market demand. Gas turbines, for instance, afford such flexibility. The premature scale up of unit size can also prove costly (Wilson 2012). Producing at smaller scales for different locations offers even more opportunities for learning in different contexts. On the other hand, sometimes scaling up unit size can supply advantages, such as improved energy efficiency. One example of the latter is the 35 percent gain in energy efficiency associated with moving from subcritical pulverized coal technology to ultra-supercritical coal technology (Beér 2009). In summary, market scale is independent from unit scale because policy-induced demand for cleaner energy technologies could be met through a diversity of energy technologies, small or large.

In fact, Schumacher (1973) was a proponent of small-scale units for the "convenience, humanity, and manageability of smallness," but also recognized the "duality of the human requirement when it comes to size," acknowledging that for different purposes, people need different structures, both small and large. Small- and large-scale cleaner energy technologies will be required to meet heterogeneous demand.

Bigger markets are certainly beautiful to producers of technology. A large market induces producers to compete for a share of it, and this competition can, in turn, drive down costs. From the consumers' point of view, a greater number of suppliers in the market (whatever country they come from) creates choice about which products to buy to best meet their needs.

Clean Energy Innovation Is No Longer a National Process

Diffusion is part of the energy technology innovation system and strongly depends on market formation by governments. Clean energy technologies are no longer developed and deployed within a single country. This may have been the situation in the 1960s and 1970s, when wind development and deployment occurred in Denmark (Karnoe and Buchhorn 2008), and solar development and deployment took place in the United States (especially in California) (see, for example, Sawin 2001; Taylor 2008), but it is no longer the case as of 2012. Also, because technological development is cumulative, iterative, and derives spillovers, countries like China are able to pursue "catch-up" strategies to acquire technological knowledge. Often these catch-up strategies require extensive imports of foreign technologies through licenses, joint ventures, or other means. These technologies are then adapted to the local market and eventually may be able to be exported. While national innovation systems are still critical to countries' abilities to enhance their absorptive capacities, evidence from these case studies shows that the Chinese have acquired technologies from all over the world, found ways to add value or reduce costs, and discovered ways to take advantage of foreign markets.

China's emergence as a global player in cleaner energy markets as a source of both demand and supply directly contributed to the globalization of these industries in several ways. Chinese studied science and engineering at home, and then went overseas to get educated technologically, and in so doing acquired a global perspective. Some of these Chinese students stayed and contributed to foreign firms, such as China's current minister of science and technology who, as mentioned earlier, worked at Audi in Germany for many years. Others took the knowledge they gained abroad and employed it in new industries in China. In the space of a decade, China became a major player in global trade in cleaner energy technologies, as discussed below.

China's clean energy manufacturers have also contributed to remarkable reductions in the global cost of these technologies. Some have argued that these cost reductions are the result of illegal subsidization in China, but illegal or not, the costs have undoubtedly fallen at a rate that nobody expected. These technologies are consequently much more accessible globally than they were only a few years ago. In other words, Chinese manufacturers have shifted the global supply curve, which has led to growth in global demand. Domestically, the Chinese clean energy market has grown 37 percent over the past five years and accounted for 20 percent of G-20 clean energy investment in 2011 (Pew Charitable Trusts 2012). China has the largest renewable energy capacity in the world (with and without its hydro capacity), with the largest solar hot water capacity, wind power capacity, geothermal direct heat use, and hydro capacity in the world (Sawin 2012).

The Costs of Cleaner Energy Technologies Are a Challenge, But Are Usually Not the Major Barrier to Their International Diffusion If Market-Formation Policies Exist

In terms of pure technological diffusion, because the costs of many cleaner energy technologies have fallen so dramatically, cost is no longer the barrier it once was (see, for example, Neij 1999; Lewis 2007). If environmental and health damages were priced into the costs of conventional technologies, many of the cleaner technologies would now be able to easily compete. One estimate of the global levelized cost of electricity as of 2012 shows that already onshore wind, small and large hydro, and geothermal are cost competitive with NGCC power plants, and that if a modest tax on carbon were imposed on fossil fuels, biomass gasification, biomass incineration, and solar PV would almost be competitive as well. The selling prices of PV cells fell 75 percent from their 2008 levels by the end of 2011. The cost of onshore wind fell 25 percent between 2009 and 2011 (McCrone 2012). Costs are context dependent, however, and there are surely some places where the costs are higher and more of a barrier than the global picture would suggest—for instance, areas with poor solar insolation.

Many of the experts interviewed during the course of my research identified cost as a barrier in the absence of policy. When asked to rank the barriers, many struggled to decide between cost and inadequate policies because they are intimately connected. The reason government is justified in intervening in the market is to correct for the market externalities. Now that many governments have made these interventions through the use of feed-in tariffs, carbon taxes, renewable portfolio standards, and other policies, cleaner energy technologies are either cost competitive or almost there.

Because of the recent cost reductions, demand has grown for the cleaner technologies, and this has a positive reinforcing effect on the dynamic created by those market-formation policies. Indeed, as costs come down, the need for the market-formation policies recedes, and the policies can be scaled back as some European countries are already doing with their feed-in tariff reductions.

From the point of view of localization, the cost of licensing or otherwise acquiring foreign technologies is a problem in some cases, but not all. In the cases in this book, the costs of foreign technologies were deemed (at certain historical points) by the Chinese to be too high for gas turbines, advanced batteries, and coal gasification. Chinese solar manufacturers made no such complaints. Still, these costs did not present overwhelming challenges for coal gasification development. In gas turbines, none of the foreign manufacturers said they were willing to license their latest (and most energy-efficient) turbine at any price.

Costs are still a serious problem in the case of advanced batteries, as is true with many new technologies until further innovation is done. These costs do not appear to be unique to China or any other country; they are simply a function of the ongoing technical challenges of adapting advanced batteries for use in automobiles.

Beyond Niche Markets

The role of niche markets is discussed in detail in the next section. Suffice it to say here, though, that the widespread global diffusion of cleaner energy technologies requires government policy that goes well beyond the creation of strategic niche markets (Kemp, Schot, and Hoogma 1998). A central argument of this book is that government-induced market formation for cleaner and more efficient energy technologies is essential because of the need to correct for market externalities, including climate change, air and water pollution (and their impact on public health), and energy insecurity. Niche markets are by definition too narrow to achieve the scale of the market that is required.

An Integrated Theory of the Diffusion of Clean Energy Innovations

The global diffusion of cleaner and more efficient energy technologies is a key process in the global energy innovation system (Grübler et al.

2012). The global energy technology innovation system "comprises all aspects of energy transformations (supply and demand); all stages of the technology development cycle; as well as all the major innovation processes, feedbacks, actors, institutions, and networks" (Gallagher et al. 2012). The global diffusion of cleaner energy technologies can also be considered integral to technological transition in the energy sector.

International technology transfer occurs through many mechanisms, as described in chapter 1. Most diffusion happens through markets where private sector firms, entrepreneurs, and consultants enter into technology-sharing agreements with one another. Governments or intergovernmental organizations, most notably the Global Environment Facility, conduct a small fraction of clean energy technology transfer. The global diffusion of cleaner and more efficient energy technologies is required to reduce the threat of global climate change, and such diffusion can also improve local environmental conditions and enhance national energy security. International technology transfer is also required if firms or countries wish to develop industrial capacity in a given technology or sector, but localization is not required to achieve the deployment of cleaner energy technologies in a given country. In this section, an integrated theory of technology transfer for cleaner and more efficient energy technologies is developed. This theory takes into account the findings in this book as well as the key distinction between technology transfer for *global diffusion* versus technology transfer for *industrial development*.

In general, technology diffusion begins in a highly localized fashion—at the core—and then starts to diffuse throughout a region and eventually the world—the periphery. The rate and timing of the diffusion process varies, but the periphery tends to have faster adoption rates because it can take advantage of the learning and experience gained in the core area (Grübler 1998). The diffusion of most technologies usually follows an S-shaped pattern with slow growth in the beginning, followed by accelerating growth, and a deceleration as saturation of the market occurs. Technology diffusion benefits from clustering and spillovers in particular locations, and also benefits from the coevolution of technologies (Grübler et al. 1999). Barriers and incentives for the cross-border transfer of technologies *in general* have been exhaustively analyzed, and two barriers are found to be particularly challenging: technological "regimes" that clearly reinforce dominant technologies or technological systems, making it difficult for new technologies to overcome the resistance of incumbents (Abernathy and Utterback 1978; Nelson and Winter 1982; Dosi 1982) and social resistance to new technologies (see, for instance, Berkhout, Smith, and Stirling 2004). It has also long been recognized that the diffusion process is affected by the communication process about the new technology (Rogers 1995).

Technologies that are transferred must also be *appropriate* for the local conditions. As previously noted, appropriate does not necessarily mean small. It means that the technology will work in the context for which it is destined, if not designed. One definition of appropriate technology is that it must be "efficient, not be obsolete, and that it must vary according to the particular situation of each country under consideration" (Reddy and Zhao 1990, 292). The original coal gasifiers licensed to China were not appropriate because they were not adapted for use with Chinese coals, which had unique characteristics. A corollary to the concept of appropriateness, however, is flexibility. Harvey Brooks (1995) defined technology transfer as "linking knowledge to need," which implies the modification of technology to specific circumstances. The more flexible a technology is for different contexts, the more likely it is that it will diffuse more rapidly and pervasively across borders, and this can be considered a design principle for cleaner energy technologies. These different contexts are strongly influenced by policy, so the more international harmonization of policy there is, the easier the standardization process becomes.

To break through barriers, a *systemic* approach to innovation is necessary. Investments into R&D and human capabilities must be coupled with learning processes, including testing, experimentation, feedbacks, and the creation and utilization of networks among key actors. Motivated entrepreneurs and government policies that are predictable, stable, and aligned are also necessary (Grübler et al. 2012; Gallagher et al. 2012).

As developing countries import and localize cleaner energy technology in their efforts to catch up technologically, we see evidence around the world of developing countries leapfrogging to cleaner energy technologies (Lewis 2007; Sauter and Watson 2008; Walz 2010; Dixon, Scheer, and Williams 2011). The solar PV and coal gasification cases in this book add, in quite different ways, to other evidence for leapfrogging in China (Wang and Kimble 2011; Fu and Zhang 2011). The solar case is one of rapid technology acquisition from abroad, coupled with strong absorptive capacities and unrivaled manufacturing execution skills that benefit strongly from the gigantic manufacturing cluster in China. The coal gasification is a more traditional yet exemplary story of technological catch-up through learning by doing as Chinese scientists experimented with coal gasifiers licensed from foreigners over decades.

The idea of clean energy leapfrogging was first proposed by José Goldemberg (1998), following Luc Soete (1985), who argued that industrializing countries can avoid the resource-intensive pattern of economic and energy development by leapfrogging to the most advanced energy technologies available rather than following the same path of conventional energy development that was forged by the highly industrialized countries. While this concept is attractive, it is clear that leapfrogging is not an automatic process because it often requires good technology absorption capacities and government intervention (Soete 1985; Gallagher 2006b; Murphy 2001). The gas turbine and advanced-battery cases in this book add to the evidence about the limits to leapfrogging because, for different reasons, there has been no leapfrogging in either technology. In the gas turbine case, government policy and support has been inconsistent and weak, technological capabilities are poor, and until recently, there was no "natural" market for the technology since natural gas was mainly used for fertilizer production. In the advanced-battery case, the technological challenges are high even for industrialized countries, policy support has been weak especially in terms of market formation, and the web of industrialized country patents has been difficult to penetrate.

A number of particular barriers and incentives have been identified for the cross-border transfer of cleaner and more efficient technologies as well. An exhaustive list was made in the Intergovernmental Panel on Climate Change special report on technology transfer in 2000, but this report did not attempt to rank the relative importance of certain factors in hindering or incentivizing the cross-border transfer of technologies, probably because of the perception that the barriers and incentives are highly context dependent. Even so, barriers to the transfer of cleaner technologies that are persistently identified include conflicting policy signals, information asymmetries, inhibiting regulations, lack of familiarity with new technologies, higher costs, inadequate access to financing, unwillingness to retire existing capital stock, and inadequate or maladapted infrastructure (Kemp, Schot, and Hoogma 1998; IPCC 2000; Brown and Sovacool 2011). All these barriers can conspire to create carbon "lock-in" (Unruh 2000).

One of this book's contributions is to clarify which barriers and incentives for the global diffusion of cleaner energy technologies are most important in the Chinese context and, ideally, more generally. The existence (or lack thereof) of market-formation policies to overcome higher costs, access (or lack thereof) to low-cost capital, and sensible (or careless) business practices that facilitate (or hinder) technology transfer are all found to be highly important. One set of barriers/incentives that this research does not find significant for the cross-border diffusion of cleaner energy technologies is access to or infringement of intellectual property. These findings support the conclusion of a recent study that asserts, "First and foremost is the economy-wide market failure caused by the absence of a price on GHG emissions" (Brown and Sovacool 2011, 60). If the aim is not only to facilitate the international diffusion of cleaner energy technologies but also to import and localize them for domestic manufacturing, then the development of absorptive capacity, formulation of coherent and stable industrial policies, creation of an ETIS, and utilization of global learning networks are essential.

Although many types of policies can be instrumental in the diffusion process, evidence in this book indicates that market-formation policies are key to spurring the global diffusion of cleaner and more efficient energy technologies. This crucial stage of market formation encompasses demonstration and niche markets but occurs before "pervasive diffusion" (Grübler, Nakicenovic, and Victor 1999). These market-formation policies must be enacted in order to incorporate the benefits of the technologies that are not naturally valued by the market into their cost. In other words, these policies must somehow account for market "externalities" either by subsidizing the cost of cleaner or more efficient energy technologies or by placing a value of the external costs on conventional energy technologies. Imposition of a carbon tax, for example, makes fossil fuel energy more expensive and renders renewable energy technologies more competitive in the marketplace. Many different types of market-formation policies appear to create incentives for technologies to be transferred across borders, including cap-and-trade policies, feed-in tariffs for renewable energy, renewable portfolio standards, and fuel-efficiency or CO_2 performance standards.

Market-formation policies include, but are not limited to, policies designed to create niche markets. One might think of the technology diffusion process as beginning with demonstration and the creation of small niche markets where new technologies can be experimented with and debugged, and learning by doing and using can start. In niche markets, financial support is provided to help make a given technology more competitive, foster the scale-up of technologies in either unit size or number, and allow for the creation of a political-economic constituency for the alternative technology (Kemp, Schot, and Hoogma 1998). The idea of "strategic niche management" is to manage the transition from one sociotechnical paradigm to another through the creation of protected spaces for the development and early deployment of new technologies, with a strong focus on learning along with a bringing together of technology push and pull efforts (ibid.; Kemp 1997).

Drawbacks to strategic niche management are that particular technologies must be chosen to experiment with in the niches, they require an interested buyer, and they are by definition small. Once a technology is widely deployed, it can no longer be considered a niche technology. Yet some cleaner and more efficient energy technologies require sustained market formation by governments beyond an early deployment/niche stage due to the huge, persistent externalities associated with fossil fuel consumption. Typically, the US Department of Defense is cited as an exemplary creator of niche markets for advanced military technologies because it is capable of specifying technology needs, working with the developer to clarify those needs, and then making large purchases given its massive financial resources. Most departments and ministries of energy around the world do not have the financial resources of the US Department of Defense. This book shows that aggregated market creation by numerous countries was sufficient to stimulate substantial entrepreneurship and cross-border deployment of cleaner energy technologies beginning in the late 1990s and taking off in the mid-2000s. Sustained market-formation policy can be considered part of a sociotechnical regime that provides structure and reinforcement behind the technological transition to a more sustainable energy system (Geels and Schot 2007).

For cleaner and more efficient energy technology use, market-formation policies need not specify particular technologies, though they sometimes do (say, a feed-in tariff for wind versus solar). These policies can be technology neutral, and individual policies often need to be combined to produce the intended outcome. According to René Kemp (1997, 317), "There is no single best policy instrument to stimulate clean technology, all instruments have a role to play, depending on the context in which they are to be used." The market-formation policies must provide a clear long-term goal (such as reduced greenhouse gas emissions, emissions of mercury, or emissions of sulfur dioxide, or improved energy efficiency). These goals could be phased in over time to provide a predictable pathway toward achievement. Then a set of policy tools can be chosen to achieve the goal, such as support for RD&D, enactment of a carbon tax, and provision of a low-interest loan fund for the purchase or use of cleaner energy technologies. In almost every case, a single policy tool is likely to be insufficient. Market-formation policies will need to be changed and adapted over time. The level of government intervention may lessen as the costs of cleaner technologies decline, for example, or policies may need to become more stringent in order to achieve the desired goal. Key features of market-formation policy for cleaner energy are alignment, stability, and predictability (Grübler et al. 2012).

Market formation policies appear to accelerate the global diffusion process in cleaner energy technologies. In figure 4.1, we observed a strong correlation between the establishment of market-formation policies for cleaner energy technologies and growth in the volume of international trade, which is only one indicator of global diffusion.

Turning now to technology transfer for the development of cleaner and more efficient energy industries domestically, it is well understood that absorptive capacity is absolutely essential. Absorptive capacity is defined as the ability to exploit external knowledge; the ability to evaluate and utilize outside knowledge is "largely a function of the level of prior related knowledge" (Cohen and Levinthal 1990, 128). Prior knowledge can be gained through education, open sources, and experience. Such knowledge provides firms with the ability to understand what they lack, what they need, and how to apply technological knowledge once they obtain it in a commercially successful manner. Innovation capacities are the capacities to select, adopt, adapt, manufacture, and deploy technologies in particular contexts (Bell and Pavitt 1993; Bell 2012). Alice Amsden (2001) differentiated these absorptive capacities, placing them into different categories: production capabilities (to manufacture and adapt products), project execution capabilities (to establish or expand activities), and innovation capabilities (to create new products and processes). More absorptive capacity is required when importing more complex technology (Mowrey and Oxley 1997). Absorptive capacity not only includes the ability to know which technologies need to be acquired but also how to acquire and apply them, and therefore tacit knowledge must also be transmitted along with "embodied" technologies (Teece 1977). Acquiring firms may have these capabilities in-house, or they may acquire them too through hiring, for instance, consultants, technicians install and connect production equipment, or people with the expertise that they need.

Technological accumulation is a learning process, and the more experience with the process, the better the absorptive capabilities and higher the level of technological competence are likely to be. International technology transfer is a process of cumulative learning (Bell and Albu 1999; Brooks 1995). International technology transfer processes themselves are not automatic, nor do imports of technologies provide automatic improvements. Lack of incentives for learning can be caused by too much government protection of a firm or industry, limited competitive pressures, excessive competitive pressures, or trade policy, to name a few factors.

For cleaner energy technologies, a sizable domestic market appears to be helpful for localization. According to Joanna Lewis and Ryan Wiser (2007, 1854), "Virtually all the leading wind turbine manufacturers come from countries that have historically maintained strong policy environments for wind development." On the other hand, the China solar PV case shows that a domestic market is not necessary for technological assimilation and successful manufacturing. Localization is also not required for the deployment of cleaner energy technologies. Some have defined technology transfer as localization, but the cross-border transfer of technology for the purpose of technology diffusion occurs all the time and indeed is sufficient to achieve the basic goal of greater diffusion of cleaner energy technologies.

Networks are an essential component of the innovation system. Establishment of user-producer networks has been deemed vitally important (Lundvall 2009). Among producers, learning networks also play a big role in the development of and experimentation with new technologies, and while these networks have traditionally been more regional in nature, Lewis (2007) provides evidence of international learning networks for wind technologies. This book adds evidence that producer networks for cleaner energy technologies are increasingly *international* rather than *national* in nature, at least in the cases of solar PV and advanced batteries for vehicles.

Anticompetitive behavior and the monopolization of technology both hinder the global diffusion of cleaner technologies. When firms have a near or total monopoly over a technology, they can individually or collectively keep prices high, refuse to license, and even refuse to sell the final product. Industry consolidation can thus hinder international diffusion (Lewis 2007).

Ready access to low-cost capital appears to accelerate the global diffusion of cleaner energy technologies when market-formation policies exist. In the realm of business practice, evidence in this book confirms that business experience in other countries leads to more risk taking and more international technology transfer from both the acquiring and selling parties. Prior international experience may also lead to *faster* international technology transfer (Vernon and Davidson 1979).

Policy Implications

So far, national- or subnational-level policies have proven highly effective at spurring the development and deployment of cleaner technologies, not just within borders, but across them as well. One can conclude, then, that perhaps the most desirable form of "global" policy would be a set of national policies that, in aggregate, create global markets for cleaner energy technologies. If the policies could be harmonized over time, economic efficiency would be improved since firms could produce for a global market, standardize their products, and not need to adapt products as much to conform to national regulation and standards. Those policies would need to be appropriate for the specific institutional and other contexts of each country, however, and they could take different forms. Indeed, we already see the diversity of choice in policy instruments with some countries choosing feed-in tariffs, others going with clean portfolio standards, and still others settling on carbon taxes or cap-and-trade programs. The more that countries enact policies, the larger the global market, and the greater the market, the greater the incentive for the global diffusion of cleaner energy technologies. So from the point of view of spurring international technology transfer, national or subnational policymakers play a critical role.

Where China is an outlier is in its willingness and ability to finance the transition to cleaner energy technologies. Many US firms complained of difficulties obtaining financing in the United States for expansionary activities or to facilitate exports. None of the firms I interviewed in China ever mentioned these challenges. If firms in the United States have trouble financing clean energy activities, then surely many developing countries also need better financing instruments. Least developed countries will clearly need help with not only accessing finance but also paying for the incremental costs of cleaner technologies. For all these reasons, if the Green Climate Fund of the Kyoto Protocol cannot be operationalized, a specific international agreement for the provision of low-cost financing for all countries and assistance with incremental costs for least developed countries could be useful.

International policy has thus far not directly caused a large-scale diffusion of cleaner and more efficient energy technologies. The Kyoto

Protocol's CDM did lead to the transfer of some cleaner technologies to developing countries, although not on a significant scale, and it is also possible that many of these transfers would have occurred in the absence of the CDM. On the other hand, many of the national policies that were enacted do seem to have been inspired by the global climate agreements (for example, the EU Emissions Trading System and Brazil's National Climate Change Policy). This research clearly shows that there is no need to wait for an international agreement, and that national policies are both necessary and sufficient. After all, national policies would be needed to implement any international treaty anyway. A bottom-up, gradual harmonization approach could be a reasonable alternative so long as the aggregated scale is sufficient, finance mechanisms are created, and special international policies are put into place to meet the needs of least developed countries.

Conclusion

The main findings and contributions of this book are that:

- international clean energy technology diffusion is no longer a purely north-to-south phenomenon but also a north-to-south-to-north, south-south, and south-to-north process—in other words, cleaner energy technology innovation is a globalized phenomenon where developing countries, in particular China, play a major role
- market formation through government policy is a crucial component of the growth model of innovation for cleaner energy technologies because these policies create much larger market demand than would happen naturally, which in turn can help to achieve cost reductions through economies of scale in a positive feedback process
- the global diffusion of cleaner energy technologies primarily takes place through private markets, not government-organized projects
- there are no insurmountable barriers to the global diffusion of cleaner energy technologies, but the barriers that are of most concern are the lack of market-formation policies, inadequate access to low-cost capital, the incremental costs of some cleaner technologies, and unwise business practices
- access to cleaner energy technologies and infringement of clean energy intellectual property are of concern, but are not serious impediments

The conclusions are based on the existing literature and the new Chinese case studies developed in this book. A paucity of empirical case studies still exists so the new integrated theory of the global diffusion of cleaner energy technologies developed and presented in this chapter must be tested in other countries, under different conditions. Many of the case studies available to date are based in the large emerging economies (for instance, India, China, and Brazil), which from the point of view of climate mitigation is appropriate, since these countries are emerging major emitters, but in the longer term, a truly global diffusion is necessary. Because a number of countries already have characteristics similar to China's, including Brazil, South Africa, India, and Mexico, it is likely that these countries will also strongly contribute to the deepening of this globalization process. Regional clusters of countries, such as the East African Community, could combine their markets to have a global impact as well. In 2010, the East African Community established a common market for goods, labor, and capital. Indeed, some already-industrialized countries could also contribute more effectively to the global diffusion of cleaner energy technologies. Whether and how least developed countries can effectively harness these global forces is a key question worthy of serious study and experimentation (Ockwell 2012).

This book has been mainly concerned with how to accelerate the diffusion of cleaner energy technologies to mitigate climate change, improve energy security, and spur sustainable prosperity. As such, it was appropriate to focus on China as the largest overall emitter of greenhouse gases in the world. Additional case studies in China, especially on efficiency technologies, would therefore be useful, as would more focused study of the roles of provincial and local governments in industrial and market-formation policy. The development of a more nuanced understanding of China's energy-technology innovation system would also be a major contribution.

Appendix A: List of People Interviewed or Informally Consulted (Alphabetical)

List of People Interviewed or Informally Consulted (Alphabetical)

Araujo, Kathleen*	MIT and Tufts University	United States
Arulanantham, Ravi	Ex-Im Bank	United States
Berg, Wilhelm	Deutscher Industrie- und Handelskammertag	Germany
Biagini, Boni	Global Environment Facility	United States
Boehm, Hans-Peter	Siemens AG	Germany
Bracy, Dennis	China Clean Energy Forum	United States
Brady, Sabrina	American Chamber of Commerce	China
CAI Nengsheng	Tsinghua University	China
Chakravorti, Bhaskar	Fletcher School, Tufts University	United States
Chang, Jack	GE	Asia and United States
Cohen, Marc	Microsoft China (2010) and Fordham University (2011)	China
Correa, Carlos	University of Buenos Aires	Argentina
de Coninck, Heleen*	Energy Research Center of the Netherlands	Netherlands
Dirschauer, Wolfgang	Vattenfall	Germany
Duwe, Matthias	Ecologic Institute	Germany
Fajkowski, James	K&L Gates	United States
Fink, Carsten	World Intellectual Property Organization	Switzerland
Forney, Matthew	Fathom China	China
Galdiz, Isabel	Ex-Im Bank	United States
Gordon, Barton	Former US representative and K&L Gates	United States
Grübler, Arnulf*	IIASA and Yale University	Austria and United States
He Long	BYD	China
Holt, Thomas	K&L Gates	United States

Jiang Wenfeng	BYD	China
Joffe, Paul*	World Resources Institute	United States
Juma, Calestous*	Harvard Kennedy School	United States
Jung, Alexander	Vattenfall	Germany
Kammen, Daniel	World Bank and University of California	United States
Kates-Garnick, Bobbi	State of Massachusetts	United States
Kong, Bo	School of Advanced International Studies and Johns Hopkins University	United States
Krebs, Harald	Vattenfall	Germany
Kruse, Andrew	Southwest Windpower	United States
Lewis, Joanna*	Georgetown University	United States
Li Ang	WWF China	China
Li Hui	Jiayuan EV Nanjing	China
Li Wenhua	GE China Technology Center	China
Lim, H. K.	Shell	China
Lin Kai	Huaneng GreenGen	China
Liu Hengwei*	Tufts University (ca. 2010) and KAPSARC	Saudi Arabia and United States
Liu Hong	Jiayuan EV Nanjing	China
Liu Lingfang	Suntech	China
Lowe, Ed	GE Energy	United States
Lyon, Geoff	Department of Energy	China
Ma Xiaoguang	Suntech	China
Metz, Bert*	European Climate Foundation	Netherlands
Mez, Lutz	Freie Universität	Germany
Miller, Alan	International Financial Corporation	United States
Moore, Fred	Dow Chemical	United States
Moomaw, William*	Fletcher School, Tufts University	United States
Mulholland, Ryan	Department of Commerce	United States
Ni Weidou	Thermal Engineering Department, Tsinghua University	China
Noland, Kirsten	Ex-Im Bank	United States
Obertacke, Ralf	Siemens Clean Energy Center	Germany
O'Connor, Craig	Ex-Im Bank	United States
Ott, Hermann	Member of the German Bundestag	Germany
Rainer Hinrichs-Rahlwes	Bundesverband Erneuerbare Energie and European Renewables Energies Federation	Germany

Reese, Herschel	Dow Corning	United States
Ren Xiangkun	Shenhua	China
Richards, Tim	GE	United States
Roffe, Pedro	International Center for Trade and Sustainable Development	Switzerland
Ryan, Michael	George Washington University	United States
Saggi, Kamal	Vanderbilt University	United States
Schreurs, Miranda*	Freie Universität	Germany
Seligsohn, Deborah*	World Resources Institute	China
Shao Liqin	Tsinghua University	China
Shen Xi	BYD	China
Shi Zhengrong	Suntech	China
Sohn, Bruce	First Solar	United States
Song Dengyuan	Yingli	China
Sonntag, Mathias	Siemens Fuel Gasification	Germany
Stefan Kratz	Q-Cells	Germany and United States
Su Jun*	School of Public Policy and Management, and Tsinghua University	China
Vieau, David	A123 Systems	United States
Vogel, Claudia	German Energy Agency	Germany
Watson, Jim*	Sussex Energy Group, University of Sussex	United Kingdom
Wen Zongkong	Guangdong EV Project	China
Wilder, Richard	Microsoft	United States
Wu Zhixin	Qingming Electric Vehicle Co.	China
Xiao Yunhan	Chinese Academy of Science	China
Xu Honghua	Institute for Electrical Engineering and Chinese Academy of Sciences	China
Xuan Xiaowei	Development Research Center, State Council	China
Xue Lan*	School of Public Policy and Management, Tsinghua University	China
Yao Qiang	Thermal Engineering Department, Tsinghua University	China
Zhang Fang*	Tsinghua University	China
Zhang Jianfu	Huaneng GreenGen	China
Zhang Jianwei	China Automotive Technology and Research Center	China
Zhang Xiliang	Tsinghua University	China
Zhang Zhihong	World Bank Climate Investment Funds	United States

Zhao Lifeng	Chinese Academy of Science	China
Zheng Fangneng	Ministry of Science and Technology	China
Zhou Yigong	Shanghai Electric	China
Zou Yufeng	Lishen Battery Co.	China

Note: * Informally consulted

Appendix B: Case Studies of Cleaner and More Efficient Technology Transfer

Case Studies of Cleaner and More Efficient Technology Transfer

Peer-reviewed case studies (alphabetical)

Title	Brief description	Countries	Reference
Transfer of sustainable energy technology to developing countries as a means of reducing emissions: The case of Bangladesh	Focus on biomass, solar and wind, and barriers assessed are related to capital investment and complexity of technology	Bangladesh	Mohammad 2001
Transferring gas turbine technologies to the former Soviet Union: Opportunities and problems	Identifies main constraints for transferring gas turbines to the former Soviet Union	Russia and other former Soviet states	Auer 1993
Promoting energy-efficient products: Global Environment Facility experience and lessons for market transformation in developing countries	Analyzes eight projects funded by the Global Environment Facility with the goal of producing market transformation through the use of demand side management; technologies were lighting, refrigerators, chillers, and industrial boilers; institutional and regulatory changes are the most important project outcomes leading to sustained market transformation	Argentina, China, Czech Republic, Hungary, Latvia, Mexico, Peru, the Philippines, Poland, South Africa, and Thailand	Birner and Martinot 2005

Peer-reviewed case studies (alphabetical)

Title	Brief description	Countries	Reference
Clean coal technology development in China	Development of advanced coal technology	China	Chen and Xu 2010
Innovation and international technology transfer: The case of the Chinese PV industry	Chinese producers have acquired technologies through two main channels: the purchasing of manufacturing equipment and the recruitment of skilled executives from the Chinese diaspora	China	de la Tour, Galachant, and Ménière 2011
Climate change investment and technology transfer in Southeast Asia	Argument: integrating technology transfer with international investment offers a powerful way to overcome disagreements in the climate change negotiations	Indonesia, Philippines, Thailand, Vietnam	Forsyth 2003
Technology transfer, indigenous innovation, and leapfrogging in green technology: The solar PV industry in China and India	Explores leapfrogging in solar PV in China and India	China and India	Fu and Zhang 2011
China shifts gears: Automakers, oil, pollution, and development	Examines development of automobile industry in China in three joint ventures and transfer of cleaner vehicle technologies; clean techs not transferred until policies were implemented	China	Gallagher 2006a

Peer-reviewed case studies (alphabetical)

Title	Brief description	Countries	Reference
An empirical case study of the transfer of greenhouse gas mitigation technologies from annex I countries to Malaysia under the CDM	In the thirteen cases studied, the author concludes that only in one case was the CDM mechanism useful for tech transfer	Malaysia	Hansen 2011
Project-based market transformation in developing countries and international technology transfer: The case of the Global Environment Facility and solar PV	Solar PV Global Environment Facility project in India that used a market-transformation approach, but insufficient demand was created; more attention was needed for enhancing capabilities	India	Haum 2012
Interactive learning or technology transfer as a way to catch up? Analyzing the wind energy industry in Denmark and India	The development of the wind industry in India is best explained through interactive learning and not simple technology transfer	Denmark and India	Kristinsson and Rao 2008
Technology acquisition and innovation in the developing world: Wind turbine development in China and India	Comparative study of technology acquisition in wind technology	China and India	Lewis 2007
Building a national wind turbine industry: Experiences from China, India, and South Korea	Industrialization process for wind industry	China, India, and South Korea	Lewis 2011

Peer-reviewed case studies (alphabetical)

Title	Brief description	Countries	Reference
The transfer of energy technologies in a developing country context—Toward improved practice from past successes and failures	Focus is on renewable energy technologies; in terms of barriers, focus is on affordability, maintenance, knowledge, ownership, and the ability to adapt	Sub-Saharan Africa	Mabuza, Brent and Mapako 2007
International technology transfer for climate change mitigation, and the cases of Russia and China	Motivations for technology acquisition differ, capacity building important, and role of market key but increasing privatization not sufficient	China and Russia	Martinot, Sinton, and Haddad 1997
Cross-border transfer of climate change mitigation technologies: The case of wind energy from Denmark and Germany to India	Analyzes reasons for success in wind as an interactive process, including investment incentives, industry/market dynamics, technology provider characteristics, regulations, etc.	India	Mizuno 2007
Making the energy transition in rural East Africa: Is leapfrogging an alternative?	Ability to absorb is key barrier; must have capacity to adapt to new technologies, take economic risks, and modify behavior	East Africa	Murphy 2001
Learning through the international joint venture: Lessons from the experience of China's automotive sector	Focus is on the extent to which joint venture is good vehicle for technology transfer and learning	China	Nam 2011
Behind the development of technology: The transition of innovation modes in China's wind turbine manufacturing industry	Development of wind industry in China, concludes that public policy was key driving force	China	Ru et al. 2012

Peer-reviewed case studies (alphabetical)

Title	Brief description	Countries	Reference
Technology transfer of energy-efficient technologies among small and medium enterprises in India	TERI-organized technology transfer project to bring more efficient technology to small-scale foundries and glass industries; incremental changes found to be better; need for someone to identify, select, and adapt the technology for local purpose	India	Pal and Sethi 2012
Can CDM bring technology transfer to China? An empirical study of technology transfer in China's CDM projects	Study of factors affecting technology transfer in CDM in China; the proportion of total income derived from the certified emissions reductions plays a key role in the project owners' decision to adopt foreign technology	China	Wang 2010
Leapfrogging to electric vehicles: Patterns and scenarios for China's automobile industry	Explores opportunities and scenarios for leapfrogging in automotive technology	China	Wang and Kimble 2001
Technology transfer and the development of China's offshore oil industry	Barriers are language, education, cultural differences, management differences, and technology gaps; learning by changing is key; government's bargaining power with industry is critical	China	Warhurst 1991

Peer-reviewed case studies (alphabetical)

Title	Brief description	Countries	Reference
The issue of technology transfer in the context of the Montreal Protocol: Case study of India	If the alternative technology is easily accessible, commercially viable, and not covered by intellectual property rights, the transition away from ozone-depleting substances is smooth; if the technologies are protected with intellectual property rights with only a few technology owners, the transition away from ozone-depleting substances is difficult	India	Watal 2007
Selection environments, flexibility, and the success of the gas turbine	Focus is on the gas turbine and its technological development	Various	Watson 2004
Technology transfer for renewable energy: Overcoming barriers for developing countries	Case studies on solar home systems and biomass cogeneration	Various	Wilkins 2002
Research, development, demonstration, and early deployment policies for advanced coal technology in China	Explores the role of Chinese policy in the development of coal gasification in China	China	Zhao and Gallagher 2007

Non-peer-reviewed case studies (alphabetical)

Title	Description	Countries	Reference
South-South technology transfer addressing climate change	Introduces a South-South paradigm of technology transfer versus North-South paradigm of conventionally used	India and China for wind and solar, and Brazil for biofuels	Corvaglia 2010
Technology transfer and barriers in Mongolia	Barriers focused on market imperfections, institutional barriers, and information barriers	Mongolia	Dagvadorj 2006
Barriers and drivers for the deployment of renewable energy technology in developing countries case study: Solar water pumps in Ghana	Solar water pumps	Ghana	Dittmar 2008
Promoting tech transfer and deployment for renewable energy and energy efficiency in Ghana	Emphasizes the need to differentiate between initiation phase barriers, implementation phase barriers, and rollout drivers, with a short section on ways to overcome barriers	Ghana	Gboney 2009
Transfer of environmentally sound technologies: Case studies from the GEF climate change portfolio	Main focus is on the following: concentrating solar power, energy-efficient kilns for brick making, wind power, fuel cell bus, and innovative financing for energy efficiency; includes lessons learned section after analysis	Bangladesh, China, Egypt, and Mexico	Global Energy Facility 2010

Non-peer-reviewed case studies (alphabetical)

Title	Description	Countries	Reference
Technology without borders: Case studies of successful technology transfer	Focus on four key areas: capacity building, information sharing, and assessment; improving coordination and aid; engaging the private sector; and financing climate-friendly policies	Various	IEA 2001
Climate technology initiative industry joint seminar on successful cases of tech transfer in Asian countries	Includes lessons learned from various case studies in Asia, including the importance of involving the private sector	China, India, Myanmar, Thailand, and Vietnam	International Institute for Sustainable Development 2006
Energy innovation: driving technology competition and cooperation among US, China, India, Brazil	Most people in India and Brazil do not view problems with intellectual property rights as a major barrier to the development or adoption of low-carbon technologies	Brazil, China, and India	Levi et al. 2010
Barriers to technology transfer for climate change mitigation: The case of the Indian coal-fired thermal power industry	Some barriers discussed include a lack of incentives, surface energy balance systems, subjection to high political influence, lack of a regulatory framework, etc.	India	Monge 2000
Barriers to the adoption of renewable and energy-efficient technologies in the Vietnamese power sector	Ranks barriers by severity based on a survey of thirty-seven domestic experts in geothermal, small hydro, and advanced coal power-generation technologies	Vietnam	Nguyen et al. 2010

Non-peer-reviewed case studies (alphabetical)

Title	Description	Countries	Reference
Transfer of emerging energy technologies	General overview of technology transfer and barriers involved	Brazil, El Salvador, Honduras, India, Kenya, Philippines, and South Africa	Davidson and Teng-Zeng, 2003
Climate change in Asia and Brazil: The role of technology transfer	Focuses on eleven case studies by country	Bangladesh, Brazil, China, India, Iran, Japan, Korea, Sri Lanka, and Thailand	Pachauri and Bhandari 1994
Lessons from case studies of tech transfer of climate technologies in the Asian region	Focus on four cases in overcoming barriers: Ecofrig project, Indian brick industry, energy efficiency in China, and cogeneration technology in Thailand; barriers include information, financing, and institutional and capacity needs	China, India, and Thailand	TERI 2000
Technology transfer to China to address climate change mitigation	Analyzes whether and how climate change mitigation technology transfer occurs in China using the cases of seven technologies either at the deployment or diffusion stage	China	Ueno 2009

Non-peer-reviewed case studies (alphabetical)

Title	Description	Countries	Reference
International perspectives on clean coal technology transfer to China: Final report to the working group on trade and environment	Three case studies in: efficient coal-fired power plants, cleaner coal gasifies for fertilizer production, and improvements in industrial boiler performance	China	Watson and Oldham 2000
Low carbon technology transfer: Lessons from India and China	Technologies are diverse in stage of development, target markets, and scale, and the evidence is mixed with regard to limitations to developing country access	China and India	Watson et al. 2010
UK-China collaborative study on low carbon technology transfer: Final report to the Department of Energy and Climate Change	Case studies of energy efficiency in the cement industry, highly efficient coal-fired power generation, EVs, and offshore wind power	China	Watson et al. 2011

Appendix C: IPC Codes for Patent Analysis

IPC Codes for Patent Analysis

Technology	IPC codes
Advanced batteries	H01M10/052
Clean vehicles	B60K6/00,20; B60K16/00; B60L7/1,20,22; B60L8/00; B60L9/00; B60L11/18; B60W20/00; H02J7/00; H02K29/08; H02K49/10
Coal gasification	F02C3/28; C10J3
Electric energy storage	B60K6/28; B60W10/26; H01G9/155; H01M10/44; H01M10/46; H02J15/00; H02J3/28
Gas turbines	F02C; F23R
Geothermal	F01K; F03G4/0; F03G7/04; F24F5/00; F24J3/08; F25B30/06; H02N10/00
LEDs	F21K99/00; F21L4/02; H01L33/0–6; H01L51/50; H05B33/00
Solar power	F03G6/0; E04D13/00,18
Solar PV	C01B33/02; C23C14/14; C23C16/24; C30B29/06; F21L4/00; F21S9/03; G05F1/67; H01G9/20; H01L25/00,03,16,18; H01L27/30,142; H01L31/00, 03–07; H01L51/4; H01M14/00; H02J7/35; H02N6/00
Wind	B63B35/00; B63H13/00; E04H12/00; F03D; H02K7/18

Sources: IPC Committee of Experts 2012; Popp 2001; author's analysis of IPC codes.

Appendix D: Timeline of Major Market-Forming Policies

Timeline of Major Market-Forming Policies

1970	United States: Growing out of earlier policies, the Clean Air Act authorized the development of regulations at the federal and state level. Regulations and sets of standards limiting emissions from point and mobile sources were developed thereafter. The Environmental Protection Agency (EPA) was also created at the end of this year and tasked with implementing the content of these acts. The Clean Air Act has undergone two amendment processes in 1977 and 1990 (EPA 2012a).
1978	Philippines: Presidential Decree No. 1442 was an act to promote the exploration and development of geothermal resources. It provided for the recovery of expenses, tax and tariff duty exemptions, easy repatriation of capital investments and remittance of earnings, and entry of alien technical and specialized personnel (IEA/IRENA 2012).
	United States: This year saw the enactment of feed-in policy (Ren21 2012, 118, table R12). The Public Utility Regulatory Policies Act (PURPA) is sometimes considered the first feed-in tariff policy (Lipp 2007; Rickerson, Sawin, and Grace 2007). Among other things, PURPA required utilities to buy electricity from qualifying facilities at rates that were based on utilities' avoided costs. Determining what constituted avoided costs was left to the individual states (Hirsh 1999).[1]
	United States: The Energy Tax Act of 1978, extended and improved in 1980, and since superseded by the Energy Policy Act (EPAct) 1992 and EPAct 2005, implemented an excise tax exemption for ethanol- and methanol-blended gasoline, and granted tax credits for renewable energy equipment purchases by households and businesses. These credits were stepped according to investment level and added on to an existing investment tax credit (IEA/IRENA 2012).

1979	Denmark: Heat Supply Act was enacted, and then revised in 1990, 2000, and 2005 or 2006. It effectively empowered the ban of electric heating in new buildings "within a district heating or natural gas supply network." The minister made use of the empowerment in 1988 and again in subsequent years (ibid.).

Singapore: The country instituted a program whereby energy is generated and fed into the grid through a waste reclamation and incineration program where waste is incinerated at plants operated by the National Environment Agency (ibid).

United States: The Volumetric Ethanol Excise Tax Credit, begun in 1979, extended in 2005 through the EPAct, and ended in 2010, was implemented by the US Department of the Treasury and granted eligible ethanol blenders a tax incentive for gasoline blends above 190 proof (ibid.).

1980	Israel: System Article 9 in accordance with Planning and Building Regulation 5730–1970 mandated that all new buildings, with differentiation by building type, must have a solar water-heating system installed. These systems must conform to a variety of performance parameters and comply with Standard no. 579, with exemption for high-rise buildings, buildings in the shade, and those built for industry, work, or hospital purposes (ibid.).

Japan: The New Energy Development Organization was established, although the name and organization of the agency has since changed. It was established under the Law Concerning Promotion of Development and Introduction of Oil Alternative Energy in 1980 as a reaction to the oil crisis. During this time, Japan also instituted projects for small- and medium-sized hydropower plants and geothermal power generation development (ibid.).

United States: A program of "loan guarantees for biomass and alcohol fuels projects" was instigated through the Biomass Energy and Alcohol Fuels Act of 1980 (ibid.). This year also saw the implementation of the Wind Energy Systems act of 1980. Both this act along with the one related to biomass and alcohol fuels have since been superseded (ibid.).

1981	United States: The Economic Recovery Act of 1981 helped businesses "recover investments in solar, wind and geothermal property through depreciation deductions" under the Accelerated Cost Recovery System (ACRS), a modified version of which (MACRS) has since superseded the original policy (ibid.).

1983	Brussels: A Royal Decree helped boost the development, demonstration, and commercialization of energy-efficiency initiatives, and overlapped with various regional initiatives (ibid.).

Iowa: This US state enacted a renewable portfolio standard or a quota policy (Ren21 2012, 119, table R13).

1984	Ireland: The Business Expansion Scheme (BES) introduced this year provided an incentive for investment in certain sectors and "investments in renewable energy companies qualify for BES relief" (IEA/IRENA 2012).
1985	Republic of Korea: A system of loans was implemented and made available to eligible applicants primarily involved in R&D, the "installation of energy conservation facilities and district heat and CHP projects" (ibid.). The loans largely originated from collections on crude oil imports (ibid.).
1986	United States: MACRS replaced the 1981 ACRS and was later updated in 2008. This year also saw the implementation of the Tax Reform Act of 1986, under which the business energy tax credit was either eliminated, phased out, or extended at a certain rate for different energy systems like biomass, solar, and geothermal. "This Act also instituted the alternative minimum tax (AMT), which significantly reduced the pool of investors who could take advantage of the tax credits." The act was thereafter superseded by various other policies beginning in 2005 (ibid.).
1987	Greece: This year marked the creation of the Centre for Renewable Energy Sources "as a legal entity whose primary aim is to promote renewable applications . . . and energy efficiency" (ibid.). Republic of Korea: Introduction of the New and Renewable Energy Development and Promotion Act, which aimed to reduce the country's reliance on imported fossil fuels as well as encourage solar heating along with the generation of heat and power from waste incineration. "The act constituted the initial framework for the development of new and renewable technologies in Korea" (ibid.). This same year saw the inception of a subsidy program for renewable energy demonstration projects and technology deployment (ibid.).
1988	Portugal: Law enacted (and later revised in 1995 and 1999) that allows renewable energy independent power producers to be sold on the grid, dependent on technical qualifications. A second law laid the groundwork for procedures to license small hydropower (ibid.). United States: The Alternative Motor Fuels Act, later extended by the Automotive Fuel Economy Manufacturing Incentives for Alternative Fueled Vehicles Rule of 2004, granted "a credit of up to 1.2 mpg toward an automobile manufacturer's average fuel economy, which helps it avoid penalties of the Corporate Average Fuel Economy (CAFE) standards." It was intended to encourage manufacturers to produce alternative fuel use, or traditional and alternative fuel use vehicles (ibid).

1989	Denmark: Changes to the Electricity Supply Act mandated that power suppliers purchase power from renewable generation, and combined heat and power. Wind was "excluded as it was already regulated in detail through the Act on Utilisation of Renewable Energy Sources." The legislation was modified in 1996 to bring it into line with the EU directive to liberalize the electricity market (ibid.). Germany: Extended from 100 to 250 MW in 1991, and ended in 2006, this "100 MW Wind Programme" provided stepped grants for the installation and operation of turbines. The program promoted 1,560 turbines with a total capacity of 362 MW (ibid.).
1990	Finland: This country becomes the first to adopt a carbon tax (Sumner, Bird, and Smith 2009). Finland's policy changed over time, especially as related to natural gas and peat (SBS Staff 2012). Originally, the tax was a surcharge on fossil fuels based on the carbon content and was set at 1.12 euros per ton of CO_2 (Environmental Administration 2012). It "was applied to gasoline, diesel, light fuel and heavy fuel oil, jet fuel, aviation gasoline, coal and natural gas" (Lin and Li 2011, sec. 3.2). In 2010, the rate was up to 20 euros per ton. The tax has undergone several revisions and has featured exemptions. For instance, the rate of taxation for natural gas was less than the general rate, and from 2005 to 2010, peat was exempted altogether. The tax was revised to a significant extent in 1997 and again in January 2011. The annual total revenues from the excise and strategic stockpile fees on energy products from 2002 to 2012 (projected) can be found within the cited source (Environmental Administration 2012). Germany: A feed-in policy (Ren21 2012, 118 table R12) was promulgated in 1990, implemented in 1991, amended in 1998, and replaced in 2000 (Ren21 2012, 118, table R12; IEA/IRENA 2012).[2] In 1990 Germany also introduced the ERP-Environment and Energy-Savings Programme, since superseded, to provide loans for entities that "take suitable measures" to save energy or, later, "plan to use renewable energies" (IEA/IRENA 2012). Netherlands: Introduced and implemented in 1990, the carbon tax at first exempted certain fuels and sectors, and the entire tax was later turned from a carbon into an energy tax, "which was equally divided into energy mixed tax and carbon tax" (Lin and Li 2011, sec. 3.2; Sumner, Bird, and Smith 2009, v). It seems that the carbon tax was actually introduced to replace a previous fuel tax that was instituted in 1988. The fuel tax "has a hybrid tax base since 1990. Initially, a fixed CO_2 component was added to the initial tax base by energy content. Since 1992, the different fuels have been (more or less) taxed according to their relative energy and carbon content, each counting for 50% in the overall tax base" (Vollebergh 2008, sec. 3). A regulatory energy tax was introduced in 1996, with its tax base mainly comprised of "excises on small-scale electricity and natural gas consumption" (ibid., sec. 1). In 2007, the country "introduced a carbon-based tax on packaging, to encourage recycling" (SBS Staff 2012).

United States: Under the Clean Air Act Amendments, a sulfur dioxide emissions cap-and-trade program was established during 1990 (Metzger 2008). This year also saw the promulgation of a small ethanol producer tax credit for ethanol production. The credit was made effective in 1992 and ended in 2010 (IEA/IRENA 2012).

1991   Australia: The Resource Management Act, a policy that featured seemingly convoluted implementation, has since undergone revision, and has recently been debated, was implemented in 1991 (ibid.).

Norway: A carbon tax—or actually two taxes—was implemented during this year (Sumner, Bird, and Smith 2009, v). One tax applied to mineral products, and the other to the petroleum industry (IEA/IRENA 2012).

Philippines: A minihydro law granted certain conditions (largely tax based) to minihydro developers ibid.).

Sweden: Another Scandinavian country also implemented a carbon tax in 1991 (Sumner, Bird, and Smith 2009, v). "Sweden enacted a tax on the use of coal, oil, natural gas, petrol and aviation fuel used in domestic travel. The tax was 0.25 SEK/kg ($US100 per tonne of CO_2) and was later raised to $US150. With Sweden raising prices on fossil fuels since enacting the carbon tax, it cut its carbon pollution by 9 per cent between 1990 and 2006" (SBS Staff 2012).

Switzerland: During 1991, the Swiss enacted a feed-in policy (Ren21 2012, 118, table R12). "With the passing of the Federal Energy Decree in 1991, electricity companies were obliged to buy electricity from renewable sources produced by private companies at a fixed rate of CHF 0.15 in the annual average. The power limit (for hydropower plants only) was set at 1 MW" (IEA/IRENA 2012).

1992   Denmark: A carbon tax was implemented on top of the existing energy taxes for an effective tax burden of 13.5 euros/ton CO_2 with substantial rebates for industry. After a tax reform in 1993, the effective tax was 80 euros/ton CO_2 with substantial tax reductions for low-income households (Weir et al. 2005).

Denmark: As well, a feed-in tariff policy similar to Germany's 1990 policy was instituted (Couture et al. 2010, 9). The initial approach was apparently abandoned in 2000 in favor of an alternate framework (Couture et al. 2010, 20).

Italy: A feed-in policy was enacted during this year (Ren21 2012, 118, table R12).

Thailand: In 1992 saw the establishment of the Energy Conservation Programme (IEA/IRENA 2012).

United States: There were several actions this year, generally associated with the national EPAct of 1992:

The Federal Business Investment Tax Credit was created by the national EPAct of 1992, and later updated in 2005, 2006, 2008, and 2009.

The State and Alternative Fuel Provider Rule was put in place requiring, for example, a certain percentage of alternative fuel vehicle procurement by covered states. The rule was amended in 2005 and 2007.

Energy Efficient Mortgages facilitated energy-efficient improvements by residents.

Renewable Energy Production Incentive provided payments for qualifying facilities. The incentive was created under the EPAct of 1992 and amended later under the EPAct of 2005 (ibid.).

1993   Denmark: In 1993, Denmark enacted a feed-in policy (Ren21 2012, 118, table R12). This seems to have been the Biomass Agreement, implemented in 1993, and modified in 1997 and 2000 (ibid.).

Finland: The Wind Power Program was established, with the aim of constructing 100 MW capacity by 2005, but it only reached 82 MW by that date (ibid.).

Germany: Although it has since ended, a Full Cost Rates program was implemented in 1993, and granted a tariff to electricity from PV and led to the eventual installation of 4.5 MW total capacity. "The programmes became obsolete after the introduction of the 100,000 Roofs Programme and the enhanced remuneration according to the Renewable Energy Sources Law" (ibid.).

India: A feed-in policy was enacted (Ren21 2012, 118, table R12). In 1993–1994, India also enacted accelerated depreciation for wind farms (ibid., 68).

Japan: The New Sunshine Program, since ended, integrated three previous initiatives, and "aimed at sustainable growth and the resolution of energy and environmental problems" (IEA/IRENA 2012). Beginning in 1993 (later ending in 2000), Japan also established a program for an "effective district energy utilization system" (ibid.).

United Kingdom: A tax on retail petroleum products was introduced (SBS Staff 2012).

United States: Administered by the EPA and since updated, Environmentally Preferable Purchasing was instituted. It required federal government entities to "make environmentally friendly purchasing decisions according to various laws, regulations, and executive orders" (IEA/IRENA 2012). This same year also marked the beginning of the Clean Cities initiative to encourage cleaner transportation (ibid.).

1994   Australia: The Renewable Energy Initiative was started. It ended in 2008 (ibid.).

Canada: The Income Tax Act, since superseded, was passed, providing tax benefits for certain kinds of renewable energy assets (ibid.).

Denmark: Changes during this year to the Electricity Supply Act included environmentally sound of supply as the act's main objective. Obligations regarding fuel use, efficiency, and renewable energy development could be placed on utilities at the minister's discretion. The legislation was modified in 1996 to bring it into line with the EU directive to liberalize the electricity market (ibid.).

Greece: A feed-in policy was enacted (Ren21 2012, 118, table R12).

Japan: The Subsidy Program for Residential PV Systems was established. It ended in 2006 (IEA/IRENA 2012).

Luxembourg: A feed-in policy was enacted (Ren21 2012, 118, table R12).

Minnesota: This US state enacted a renewable portfolio standard or a quota policy (ibid., 119, table R13).

Spain: A feed-in policy was enacted (ibid.,118, table R12). "Royal Decree on electricity produced by hydro sources, cogeneration and RES (R.D. 2366/1994) setting the basic conditions of the contract between the RES-E producer and the distribution companies" (del Rio and Gual 2007, sec. 3).

Sweden: An Environmental Bonus for Wind Power was established during this year. The bonus ended in 2009 (IEA/IRENA 2012).

Thailand: Small and Very Small Power Purchase Agreements were established (ibid.).

1995   Austria: The federal energy R&D budget for this year allocated around 32 percent specifically for renewable sources. This decreased to around 25 percent the following year, but the lower figure still stood about three times higher than the IEA average (ibid.).

Denmark: The Green Tax Package was set in place during this year, later to be superseded by the Carbon Tax/Green Tax System in 1999 (ibid.).

Germany: The 100 Million Program was established. It designated 100 million DM, was administered by the Federal Ministry of Economics and Technology, and provided capital subsidies to encourage renewable energy use. The program was ended in 1998 (ibid.).

1996   Arizona: This US state enacted a renewable portfolio standard or a quota policy (Ren21 2012, 119, table R13).

Canada: The Canadian Renewable Conservation Expenses, still in force, allowed deductions for renewable energy and conservation start-up projects. The initiative was expanded in 2002 and amended in 2010 (IEA/IRENA 2012).

China: The Brightness Program was instituted. According to the IEA/ IRENA, the Brightness Program, which is still in force, "is an umbrella program, which includes the Township Electrification Program (TEP) (implemented 2002–2003), and the Village Electrification Program. The TEP included 20 MW of solar PV and wind, and 200 MW of small hydro to provide electricity for more than 1000 townships (the official statistic is 1 million people total). To date, the central government has invested USD 240 million to provide hardware. Thus, it is designed to allow development of rural communities as well as reducing poverty. Total investment in equipment and services needed to achieve the long-term goal of reaching 23 million people is about 10 billion Renminbi (about USD 1.2 billion). The government of Holland and Germany are giving further financial and technical support to the programme" (ibid.).

France: The Wind Energy Program, dubbed the "EOLE" Program, was intended to facilitate the establishment of at least 250 MW of grid-connected wind energy by 2005 (ibid.).

Germany: In 1996, Germany began introducing "green electricity" as a product on the market. The electricity was derived from "renewable energy plants not operating under the German Feed-In Scheme, the EEG" (ibid.).

United States: The State Energy Program consolidated earlier programs and is still in force. This program aids states in their own renewable energy and efficiency initiatives through grants (ibid.).

1997   Canada: Government purchases of electricity from renewable resources were implemented in 1997, first effective in 2001, and ended in 2007. The aims of this initiative included displacing emissions, providing a "first customer" to aid utilities in acquiring familiarity with new products, and creating "viable green power markets" (ibid.). Also this year (and also later ended in 2007), Canada announced a Renewable Energy Deployment Initiative with funding directed to three areas: "market stimulation, industry infrastructure support, and market development" along with additional training and support activities (ibid.).

Costa Rica: A tax on carbon pollution was enacted, "set at 3.5 per cent of the market value of fossil fuels. The revenue raised from this goes into a national forest fund which pays indigenous communities for protecting the forests around them" (SBS Staff 2012).

Denmark: A tax incentive for wind was established (IEA/IRENA 2012).

Japan: Several initiatives were implemented this year, including a program to develop and disseminate PV systems, a law to accelerate the advancement of the introduction of new energy, subsidies supporting new energy, and a program to support the deployment of new and renewable energy (ibid.).

Maine, Massachusetts, and Nevada: These US states enacted renewable portfolio standards or quota policies (Ren21 2012, 119, table R13).

Philippines: Executive Order 462: New and Renewable Energy Program was signed into law in 1997, and modified in 2000. The program encompassed several aims, such as increasing the role of the private sector in developing new and renewable energy (IEA/IRENA 2012).

Spain: A feed-in tariff policy similar to Germany's 1990 policy was enacted (Couture et al. 2010, 10). The Law on the Electricity Sector established tariffs at 80–90 percent of the retail rate, but the framework was replaced in 1998 with RD 2818/1998 and again in 2004 with a percentage-based design. Finally, this third policy was abandoned in favor of offering the option of fixed prices and sliding premiums (Couture et al. 2010, 20).

Sri Lanka: A feed-in policy was enacted (Ren21 2012, 118, table R12).

Sweden: The Energy Policy Program was established with the aim of making renewables more competitive with nuclear power and fossil fuels. The program "included a seven-year RD&D programme of SEK5.6 billion (Eur 93 million per year) for renewable energy sources and new energy technology" (IEA/IRENA 2012). This year also saw the implementation of a power purchase initiative, an RD&D program for renewable energy sources and new energy technology, and grants for the increased use of renewable energy sources (ibid.).

1998   Connecticut, Pennsylvania, and Wisconsin: These US states enacted renewable portfolio standards or quota policies (Ren21 2012, 119, table R13).

Japan: The Top Runner Efficiency Program was established to promote energy efficiency in appliances and vehicles through the imposition of efficiency standards. Twenty-one items (e.g., computers, refrigerators, and televisions) were regulated (Lau et al. 2009).

Sweden: A feed-in policy was enacted (Ren21 2012, 118, table R12).

1999   Denmark: Subsidies, since ended, for wind turbines were established (IEA/IRENA 2012).

Germany: A Market Incentive Program, or Marktanreizprogramm, was put into place. This program followed the "100 Million Program" and theoretically triggered billions of euros in investment in the years 2007–2010 (ibid.).

Italy: This year saw the enactment of a renewable portfolio standard or quota policy (Ren21 2012, 119, table R13).

New Jersey and Texas: These US states enacted renewable portfolio standards or quota policies (ibid., 119, table R13).

Norway: A feed-in policy was enacted (ibid., 118, table R12).

Portugal: A feed-in policy was enacted (ibid., 118, table R12).

Slovenia: A feed-in policy was enacted (ibid., 118, table R12). United States: The Alternative Fuels and Fleet Efficiency Program along with the Wind Powering America initiative were both implemented during 1999. The wind program was intended to "facilitate the installation of at least 100 MW of wind in at least 30 States by 2010" (IEA/IRENA 2012). This year also marked the reestablishment of a production tax credit through the US Ticket to Work and Work Incentives Improvement Act of 1999.

2000   New Mexico: This US state enacted a renewable portfolio standard or quota policy (Ren21 2012, 119, table R13).

France: Electricity Law 2000 addressed the "obligatory purchase of electricity from renewable sources and cogeneration at fixed feed-in tariffs" (IEA/IRENA 2012).

Germany: The Renewable Energy Sources Act (Erneuerbare Energien Gesetz, EEG) was adopted and replaced the Electricity Feed-In Law of 1991. It was aimed at doubling "the share of electricity produced from renewable energy by 2010" (ibid.). There was also a national-level decoupling of feed-in tariff prices from electricity prices and several other major developments (Couture et al. 2010, 10).

United Kingdom: The Renewables Obligation Plan was established. Since superseded, and part of the Utilities Act 2000, the plan "require[d] licensed electricity suppliers to buy specified portions of their purchases . . . from renewable sources" and was intended to last for twenty-five years (IEA/IRENA 2012).

United States: This year saw the implementation of programs to enhance biomass and bioenergy RD&D (ibid.).

2001   Armenia: A feed-in policy was enacted (Ren21 2012, 118, table R12). Australia: A renewable portfolio standard or quota policy was enacted (ibid., 119, table R13). This policy was likely the Mandatory Renewable Energy Target, which aimed to integrate renewable energy into the electricity mix up to a per year contribution of 9,500 GWh by 2010. Tradable Renewable Energy Certificates "are used to demonstrate compliance with the objective" (IEA/IRENA 2012).

Canada: Sustainable Technology Development Canada was set up as a fund for the development and demonstration of "environmental technologies, particularly those aimed at reducing greenhouse gas emissions and improving air quality" (ibid.). The funding has since reached 1.5 billion CAD (ibid.).

China: A Reduced Value Added Tax was established for wind, power generation from municipal solid waste, and later biogas (ibid.).

European Union: The Directive on Electricity Production from RE Sources was signed into law, setting national indicative targets for renewable energy production from individual member states (Europa 2012).

Flanders: This semiautonomous region of Belgium enacted a renewable portfolio standard or quota policy (Ren21 2012, 119, table R13).

France: A feed-in policy was enacted (Ren21 2012, 118, table R12). Established under the Electricity Law of 2000, these feed-in tariffs affected wind energy, small hydro, combustible waste, solar, biogas from landfills, municipal solid waste (other than biogas), and cogeneration (IEA/IRENA 2012).

Italy: A Grant for Solar Thermal Research was set up, although it ended in 2003. This initiative charged the National Agency for New Technology, Energy, and Environment with RD&D on solar thermal generation, and funded this work through the 2001 Italian budget law to the tune of two hundred billion euros over three years (ibid.).

Latvia: A feed-in policy was enacted (Ren21 2012, 118, table R12).

Republic of Korea: The Energy RD&D Basic Plan was established. An update of a 1987 program, this initiative aimed to invest approximately eight hundred million US dollars from 2001 to 2006 through tax benefits, renewable portfolio standards, and other measures. Wind and PV were deemed top priorities (IEA/IRENA 2012).

United Kingdom: The Climate Change Levy began in this year (Sumner, Bird, and Smith 2009, 1), "designed to promote energy efficiency and stimulate investment in new energy technologies" (IEA/IRENA 2012).

2002   Algeria: A feed-in policy was enacted (Ren21 2012, 118, table R12).

Australia: Ethanol and Biodiesel Production Grants were set up (IEA/IRENA 2012).

Austria: A feed-in policy was enacted (Ren21 2012, 118, table R12), likely the Green Electricity Act, which entered into force on January 1, 2003 (IEA/IRENA 2012).

Brazil: A feed-in policy was enacted (Ren21 2012, 118, table R12). This is likely the passage of Law 10438, which led to the PROINFA program, implemented in two stages (IEA/IRENA 2012).

California: This US state enacted a renewable portfolio standard or quota policy (Ren21 2012, 119, table R13). "Established in 2002 under Senate Bill 1078, California's Renewables Portfolio Standard (RPS) was accelerated in 2006 under Senate Bill 107 by requiring that 20 percent of electricity retail sales be served by renewable energy resources by 2010. Subsequent recommendations in California energy policy reports advocated a goal of 33 percent by 2020, and on November 17, 2008, Governor Arnold Schwarzenegger signed Executive Order S-14-08 requiring that . . . [a]ll retail sellers of electricity shall serve 33 percent of their load with renewable energy by 2020." The following year, Executive Order S-21–09 directed the California Air Resources Board, under its Assembly Bill 32 authority, to enact regulations to achieve the goal of 33 percent renewables by 2020. Senate Bill X1–2, signed into law by Governor Jerry Brown in 2011, superseded this order (California Energy Commission 2012b).

Canada: The Wind Power Production Incentive, ended in 2007, was launched with the expectation of expanding the country's installed wind capacity by 500 percent over the course of five years, and spending reached 260 CAD million by the end of March 2007 (IEA/IRENA 2012).

Czech Republic: A feed-in policy was enacted (Ren21 2012, 118, table R12).

France: A feed-in tariff was established to support projects under twelve MW. The covered sources were biomass, methanization, geothermal, animal waste, and Solar PV (IEA/IRENA 2012).

Indonesia: A feed-in policy was enacted (Ren21 2012, 118, table R12). This policy was set for renewable energy plants at or below 1 MW, and was later expanded in 2006 to accommodate larger production. A 2009 update revised pricing regulation (FuturePolicy 2011).

Israel: Initially, the Renewable Energy Targets were set at getting 2 percent of electricity from renewable energy by 2007, but that target was later shifted (IEA/IRENA 2012).

Lithuania: A feed-in policy was enacted (Ren21 2012, 118, table R12).

United Kingdom: A renewable portfolio standard or quota policy was enacted (ibid., 119, table R13). The Renewables Obligation, which started out largely technology neutral, was amended in 2009, follows from the Non Fossil Fuels Obligation, and was "the Government's main support mechanism for renewables" (IEA/IRENA 2012). This year also saw an allocation for the Major PV Demonstration Program, a reduction in taxes on biofuels, the provision of capital grants for offshore wind, and the launch of the Bio-energy Capital Grants Scheme (ibid.).

United States: The Job Creation and Worker Assistance Act of 2002 reestablished production tax credits.

Wallonia: This region of Belgium enacted a renewable portfolio standard or quota policy (Ren21 2012, 119 table R13).

2003   China: Preferential tax policies for renewable energy were instituted during this year, including a reduction in the tax rates for foreign investment in biogas and wind energy production. These initiatives were later expanded in 2007. A Wind Power Concession Program also ran from 2003–2009, and domestic and international companies bid for 100–200 MW projects. Concessions initially featured domestic content requirements (IEA/IRENA 2012).

Cyprus: A feed-in policy was enacted (Ren21 2012, 118, table R12).

Estonia: A feed-in policy was enacted (ibid., 118, table R12).

Hungary: A feed-in policy was enacted (ibid., 118, table R12).

Indonesia: Geothermal Law 27/2003 was enacted and aimed partly to help facilitate upstream developments, including from the private sector (IEA/IRENA 2012).

Japan: A renewable portfolio standard or quota policy was enacted (Ren21 2012, 119, table R13).

Maharashtra: This Indian state enacted a feed-in policy as well as a renewable portfolio standard or quota policy (ibid., 118, table R12; ibid., 119, table R13).

Slovak Republic: A feed-in policy was enacted (ibid., 118 table R12).

South Korea (Republic of Korea): A feed-in policy was enacted (ibid., 118, table R12).

Sweden: A renewable portfolio standard or quota policy was enacted (ibid., 119, table R13).

United Kingdom: The Renewable Energy Guarantee of Origin was established (IEA/IRENA 2012).

2004   Andhra Pradesh, Karnataka, Madhya Pradesh, and Orissa: These Indian states enacted renewable portfolio standards or quota policies (Ren21 2012, 119, table R13).

Andhra Pradesh and Madhya Pradesh: These Indian states enacted feed-in policies (ibid., 118, table R12).

Colorado, Hawaii, Maryland, New York, and Rhode Island: These US states enacted renewable portfolio standards or quota policies (ibid., 119, table R13).

Finland: The ClimBus Energy Program, ended in 2008, was partially aimed at enhancing "Finnish companies as internationally important suppliers of technology and services related to climate change mitigation," and was funded in excess of ninety million euros (IEA/IRENA 2012).

Israel: A feed-in policy is enacted (Ren21 2012, 118, table R12). This initiative seems to have applied to generation from wind and PV (IEA/IRENA 2012).Nicaragua: A feed-in policy was enacted (Ren21 2012, 118, table R12).

Nova Scotia, Ontario, and Prince Edward Island: These Canadian provinces enacted renewable portfolio standards or quota policies (ibid., 119, table R13).

Poland: A renewable portfolio standard or quota policy was enacted (ibid., 119, table R13).

Prince Edward Island: This Canadian province enacted a feed-in policy (ibid., 118, table R12).

Spain: Since superseded, the Special Regime for the Production of Electricity from renewable energy sources (Royal Decree 436/2004) amended the earlier Royal Decree 2818/1998 (IEA/IRENA 2012).

Thailand: This year saw the release of an 8 percent target (of commercial primary energy) to be met through small power producers and very small power producers programs, a renewable portfolio standard, and other incentive programs and funding (ibid.). United Kingdom: The Energy Act 2004 was established (ibid.). United States: The Hydrogen Program, administered by the US Department of Energy and involving several of its offices, was introduced. This year also saw the extension of a production tax credit with planned funding of ten billion dollars from 2007 to 2013 (ibid.). This extension took place through the Working Families Tax Relief Act of 2004.

2005   China: A feed-in policy for wind energy is enacted through the Renewable Energy Law, which came into effect January 1, 2006. The Renewable Energy Law, revised in 2009, was a framework policy and set out a path for the integration of renewable energy into China's energy mix. It included a renewable energy target along with a mandatory connection and purchase policy (ibid.). Also implemented this year, the eleventh five-year plan included several targets related to renewable energy as well as an energy intensity target (ibid.; Ren21 2012, 118, table R12).

Delaware, District of Columbia, and Montana: These US states enacted renewable portfolio standards or quota policies (Ren21 2012, 119, table R13).

Ecuador: A feed-in policy was enacted (ibid., 118, table R12).

Germany: The Fifth Energy Research Program, or 5. Energieforschungsprogramme—Innovation und Neue Energietechnologien, was established. It replaced the Fourth Energy Research Program started in 1996. Also from 2005 to July 2006, Germany disbursed 784 euros in loans for small investments in solar PV generation (IEA/IRENA 2012).

Gujarat: This Indian state enacted a renewable portfolio standard or quota policy (Ren21 2012, 119, table R13).

Indonesia: In 2005, the country implemented the National Energy Blueprint, which (in part) set targets for specific sources of renewable energy such as geothermal, small-scale hydropower, solar energy, biomass for power generation, and wind energy (IEA/IRENA 2012).

Ireland: A feed-in policy was enacted (Ren21 2012, 118, table R12).

Italy: The Ministry of Economic Development issued decrees during the year, and in February 2006 introduced "a feed-in premium scheme for solar PV" (ibid.). The scheme has since been revised (ibid.).

Karnataka, Uttaranchal, and Uttar Pradesh: These Indian states enacted feed-in policies (Ren21 2012, 118, table R12).

Norway: The Domestic Emissions Trading Scheme ran from 2005 to 2007 (IEA/IRENA 2012).

Spain: The Renewable Energy Plan (Plan de Energías Renovables en España, PER) was instituted for 2005–2010. This plan revised the earlier 2000–2010 plan, and among its provisions (including specific targets for electricity, transport fuel, and MW for specific technologies), aimed at attaining at least 12 percent of the total energy use from renewable sources by 2010 (ibid.).

Turkey: A feed-in policy was enacted (Ren21 2012, 118, table R12).

United States: The Renewable Fuel Standard was created under the EPAct of 2005. The Renewable Fuel Standard 2 has since adjusted the first one (EPA 2012b). The EPAct of 2005 as a whole was instituted this year. The Renewable Fuel Standard Program eventually took effect in September of 2007 (IEA/IRENA 2012). The act of 2005 reestablished production tax credits and included new loan guarantees for low carbon energy sources.

2006   Argentina: A feed-in policy was enacted (Ren21 2012, 118, table R12).

California: This US state passed the Global Warming Solutions Act of 2006, which required the development of measures to reduce greenhouse gas emissions by 2020.

France: The Renewable Energy Feed-In Tariffs (III) was enacted (IEA/IRENA 2012).

Indonesia: The National Biofuel Roadmap 2006–2025 was implemented (ibid.).

Kerala: This Indian state enacted a feed-in policy (Ren21 2012, 118, table R12).

Ontario: This Canadian province enacted a feed-in policy (ibid., 118, table R12).

Pakistan: A feed-in policy was enacted (ibid., 118, table R12).

Thailand: A feed-in policy was enacted (ibid., 118, table R12). IEA/IRENA (2012) reports the implementation of a feed-in premium for renewable power in 2007, with a modification in 2009.

United Kingdom: A Microgeneration Strategy was created this year (ibid.).

United States: The Green Purchasing Affirmative Procurement Program established guidelines for the US Department of Agriculture for the purchase and use of products. Also in 2006, a residential renewable energy tax credit established within the federal EPAct of 2005 came into effect, and was later amended in 2008 and 2009. Finally, the Solar America Initiative, which was later superseded, came into effect and was funded at approximately a billion US dollars (ibid.).

Washington: This US state enacted a renewable portfolio standard or quota policy (Ren21 2012, 119, table R13).

2007   Albania: A feed-in policy was enacted (Ren21 2012, 118, table R12).
Alberta: This Canadian province introduced a carbon tax (SBS Staff 2012).
Argentina: This year saw the establishment of biofuels promotion laws (IEA/IRENA 2012).
Boulder: This city in the US state of Colorado introduced a carbon tax (Sumner, Bird, and Smith 2009, 2).
Bulgaria: A feed-in policy was enacted (Ren21 2012, 118, table R12).
California: The California Solar Initiative aimed to "put solar on a million roofs" in this US state through a ten-year program (IEA/IRENA 2012).
Canada: The ecoENERGY for Renewable Power program was implemented, with a plan to invest approximately 1.5 billion CAD. A production incentive of 1 Canadian cent per kWh within certain bounds and from April 1, 2007 to March 31, 2011 was meant to aid the contribution of renewable energy. The same year, Canada implemented several other initiatives and incentives (ibid.).
China: A renewable portfolio standard or quota policy was enacted (Ren21 2012, 119, table R13) This year also saw the release of the National Climate Change Program, signing of a memorandum of understanding by the United States and the People's Republic of China on biomass, and implementation of a wind farm plan in Hainan Province, aiming to meet interim targets on the way to situating 600 MW by 2020 (IEA/IRENA 2012).
Croatia: A feed-in policy was enacted (Ren21 2012, 118, table R12).
Dominican Republic: A feed-in policy was enacted (ibid., 118, table R12).
Egypt: The Supreme Council of Energy announced a target of 20 percent generation of electricity from renewable energy by 2020 through two planned phases: a competitive bids approach for wind, and a feed-in tariff (IEA/IRENA 2012).
Finland: A feed-in policy was enacted (Ren21 2012, 118, table R12).
Illinois, New Hampshire, North Carolina, Northern Mariana Islands, and Oregon: These four US states and one commonwealth enacted renewable portfolio standards or quota policies (ibid., 119, table R13).
Macedonia: A feed-in policy was enacted (ibid., 118, table R12).
Moldova: A feed-in policy was enacted (ibid., 118, table R12).
Mongolia: A feed-in policy was enacted (ibid., 118, table R12).
Philippines: The Biofuels Act became law, establishing local-sourcing minimums and providing various incentives. The act seemed to be directed primarily toward the domestic market (IEA/IRENA 2012).
Quebec: This Canadian province introduced a carbon tax (Sumner, Bird, and Smith 2009, 2).

South Australia: A feed-in policy was enacted (Ren21 2012, 118, table R12).

Spain: RD 661/2007 introduced a "sliding premium" option for feed-in tariffs that offered payment or premium above market, and was intended to ensure profitability. It was shortly followed by a similar policy in the Netherlands (Couture et al. 2010, 10). This decree superseded Royal Decree 436/2004 (IEA/IRENA 2012).

Uganda: A feed-in policy was enacted (Ren21 2012, 118, table R12).

United Kingdom: This year saw the creation of the Environmental Transformation Fund, which then funded 400 million GBP in the period from 2008–2009 to 2010–2011 (IEA/IRENA 2012).

United States: The US Department of Energy's Loan Guarantee Program authorized billions in loan guarantees. It was eventually partially superseded by the American Recovery and Reinvestment Act. In September of this year, the Renewable Fuel Standard Program took effect (ibid.). Additionally, the US Energy Independence and Security Act authorized new efficiency standards.

2008 Australia: The Geothermal Drilling Program was created during this year, and the first round of funding provided seven projects with seven million AUD each (ibid.).

Bay Area Air Quality Management District: This California entity implemented a carbon tax (Sumner, Bird, and Smith 2009, 2).

Brazil: The B2 biodiesel requirement came into effect (IEA/IRENA 2012).

British Columbia: This Canadian province implemented a carbon tax (Sumner, Bird, and Smith 2009, 2).

California: This US state enacted a feed-in policy (Ren21 2012, 118, table R12). Set within the Public Utilities Code, § 399.20 created the feed-in tariff program (since amended) and authorized the "purchase of up to 480MW of renewable generating capacity from renewable facilities smaller than 1.5 MW" (California Public Utilities Commission 2012).

Chhattisgarh, Gujarat, Haryana, Punjab, Rajasthan, Tamil Nadu, and West Bengal: These Indian stateseated feed-in policies (Ren21 2012, 118, table R12).

Chile: A renewable portfolio standard or quota policy was enacted (ibid., 119, table R13). The Non-Conventional Renewable Energy Law generally required energy companies of a certain size to demonstrate a certain mandated percentage from nonconventional energy sources (IEA/IRENA 2012).

China: Shandong Province's One Million Rooftops Sunshine Plan was put into place (ibid.).

France: The New Energy Technologies Demonstration Fund, which allotted four hundred million euros over four years, was created (ibid.).

India: The National Action Plan on Climate Change outlined eight "national missions," including an emphasis on solar, and implementation plans were requested by December 2008 (ibid.). Also implemented this year was an incentive based on solar power generation (ibid.). A renewable portfolio standard or quota policy was enacted as well (Ren21 2012, 119, table R13). Finally, generation-based incentives for wind power were created to promote investment in new and large wind power producers in order to achieve a target of 10,500 MW of new wind power capacity by 2012. Investors in eligible projects would receive a payment of 0.5kWh INR for ten years over and above the tariff determined for wind power by the relevant authorities (IEA/IRENA 2012).

Japan: This year saw the revision of earlier targets and specified targets for individual oil-equivalent technologies. Additionally, Japan announced the Cool Earth Energy Innovative Technology Plan (ibid.).

Kenya: A feed-in policy was enacted (Ren21 2012, 118, table R12).

Michigan, Missouri, and Ohio: These US states enacted renewable portfolio standards or quota policies (ibid., 119, table R13).

Netherlands: Feed-in tariff sliding premiums similar to Spain's RD661/2007 were adopted (Couture et al. 2010, 10).

Philippines: A feed-in policy was enacted as well as a renewable portfolio standard or quota policy (Ren21 2012, 118, table R12; ibid., 119, table R13).

Queensland: This Australian state enacted a feed-in policy (ibid., 118, table R12).

Romania: A renewable portfolio standard or quota policy was enacted (ibid., 119, table R13).

South Korea (Republic of Korea): A national carbon tax was introduced (SBS Staff 2012).

Switzerland: A system of feed-in tariffs "differentiated by technology, size and application" was established (IEA/IRENA 2012). Switzerland also introduced a carbon incentive tax that included "all fossil fuels, unless they are used for energy. Swiss companies can be exempt from the tax if they participate in the country's emissions trading system. The tax amounts to CHF 36 per metric tonne CO_2" (SBS Staff 2012).

Tanzania: A feed-in policy was enacted (Ren21 2012, 118, table R12).

Ukraine: A feed-in policy was enacted (ibid., 118, table R12).

United Kingdom: The Climate Change Act became law, as did Energy Act 2008 as well as Planning and Energy Act 2008. Additionally, the Renewable Transport Fuels Obligation, announced in 2005, came into force (IEA/IRENA 2012).

United States: The Emergency Economic Stabilization Act of 2008, through the Energy Improvement and Extension Act, included provisions that extended the production tax and investment tax credits (which would have otherwise expired at the end of 2008) for several kinds of renewable energy. It additionally allowed utilities to begin seeking investment tax credit. Over eight hundred million US dollars in bonds were authorized for financing. Moreover, the Energy Independence and Security Act of 2007 was implemented this year, and energy provisions within the National Defense Authorization Act for Fiscal Year 2009 required the Department of Defense to consider certain renewable energy for use in the armed forces. Grants for work on advanced biofuels, led by the Department of Energy, were also authorized up to five hundred million US dollars from 2008 to 2015, and additional funds were authorized for grants to establish renewable fuel infrastructure (ibid.). This year also marked the first auction of CO_2 allowances of the Regional GHG Initiative (2012) in the northeastern United States—a cap-and-trade regime for the region.

2009   Australia: The Clean Energy Initiative included four components and was funded at 5.1 billion AUD. Among the included initiatives, funding was made available through the Australian Solar Institute. By January 2011, twenty-seven projects had been funded at a total of 66.1 million AUD. Large-scale solar was also supported and funded through the Clean Energy Initiative, and the Solar Flagships Program was funded at 1.5 billion AUD. Renewable energy and low emissions coal technologies were also included in the Clean Energy Initiative (IEA/IRENA 2012).

Australian Capital Territory, New South Wales, and Victoria: This Australian territory and two states enacted feed-in policies (Ren21 2012, 118, table R12).

Brazil: The National Policy on Climate Change was signed into law. It contained an emissions reduction target and twelve sectoral plans of action (Government of Brazil 2010).

Canada: The Clean Energy Fund invested 850 million CAD in technology development and demonstration, including renewable energy and CCS for organizations registered in Canada. There was an emphasis on system, not just technology, demonstrations (IEA/IRENA 2012).

China: The Golden Sun Program for the demonstration of solar PV was established and then revised in 2011. Amendments to the 2006 Renewable Energy Law were also implemented this year (ibid.).

Denmark: The Promotion of Renewable Energy Act, among other initiatives, established "detailed feed-in tariffs for wind power, as well as other sources of renewable energy" (ibid.).

France: The Sustainability Energy Provisions within the 2009 Finance Law included incentives such as 0 percent loans for energy-efficient renovation (ibid.).

Germany: The 2009 Amendment of the Renewable Energy Sources Act, among other things, increased the feed-in tariffs for wind energy and provided other measures to stimulate wind power development. The measurements also increased or decreased feed-in tariffs on other technologies (ibid.).

Hawaii, Oregon, and Vermont: These US states enacted feed-in policies (Ren21 2012, 118, table R12).

India: The Renewable Energy Tariff Regulations were created this year and then revised in 2010 (IEA/IRENA 2012).

Israel: Feed-in tariffs for solar PV and wind-sourced power were established, and were adjusted in 2011 (ibid.).

Japan: A feed-in policy was enacted (Ren21 2012, 118, table R12). The emphasis seemed to be on solar, and the policy was later adjusted in 2011 (IEA/IRENA 2012).

Kansas: This US state enacted a renewable portfolio standard or quota policy (Ren21 2012, 119, table R13).

Kazakhstan: A feed-in policy was enacted (ibid., 118, table R12).

Mexico: The Special Program for the Use of Renewable Energy was instituted. It implemented the 2008 Renewable Energy Development and Financing for Energy transition law, and contained three main pillars: "establishing methodologies to improve renewable energy production transformation and distribution regulations and norms"; improving the market context; and supporting entrepreneurship (IEA/IRENA 2012).

Poland: The National Fund for Environmental Protection and Water Management, with funding of up to 1.5 billion euros, began supporting investment and construction activities for renewable energy as well as "high efficiency cogeneration facilities" through loans and in some cases grants (ibid.).

Republic of Korea: The One Million Green Homes Program "merged with the existing 100,000 solar roof installations" program (ibid.). This year, the South Korean Ministry of Strategy and Finance also "implemented a 50 percent import duty reduction on 31 products used in the generation of renewable energy" (ibid.). The aims within the National Energy Basic Plan for 2008–2030 (and implemented in 2009) included a target to "emerge as a major renewable products manufacturer and target to cover 20% of the global solar manufacturing market by 2030 from about 5% in 2008" (ibid.).

Serbia: A feed-in policy was enacted (Ren21 2012, 118, table R12).

South Africa: A feed-in policy was enacted (ibid., 118, table R12). Through this policy, South Africa's public utility, Eskom, was required to "purchase the output from qualifying renewable energy generators at pre-determined prices based on the levelized cost of electricity," the cost of which will be passed on to customers. The tariffs covered several renewable energies, such as wind, small hydro, landfill gas methane, and PV (IEA/IRENA 2012).

Taiwan: A feed-in policy was enacted (Ren21 2012, 118, table R12). The tariffs were adjusted in 2011. The country also implemented the Statute for Renewable Energy Development Bill, which provided for a feed-in tariff (IEA/IRENA 2012).

United Kingdom: The wide-ranging Low Carbon Industrial Strategy touched on several specific renewable energy technologies, and aimed to "ensure that British businesses and workers are equipped to maximize the opportunities and limit the costs of transitioning to a low-carbon economy" (ibid.).

United States: The American Recovery and Reinvestment Act included over US$80 billion "to support clean energy research, development, and deployment," including over US$50 billion in direct appropriations and US$30 billion "in the form of tax-based incentives" (ibid.). The Advanced Energy Manufacturing Tax Credit, established by the American Recover and Reinvestment Act of 2009, made US$2.3 billion in tax credits available and was implemented through competitive bidding (ibid.).

2010   Argentina: The GENREN program aimed to reach the target of 8 percent of total electricity from renewable sources by 2016 as articulated by Law 26 190 in 2007. It additionally formed the Fiduciary Fund for Renewable Energy, financed by the electricity tax, to grant subsidies per kWh to qualifying power, wind, geothermal, biomass, biogas, and hydro facilities (ibid.).

Austria: This year saw the enactment of Ökostromverordnung (feed-in tariffs) and Ösvo (ibid.).

Bosnia and Herzegovina: A feed-in policy was enacted (Ren21 2012, 118, table R12).

Brazil: The 2010–2019 Plan for Energy Expansion foresaw "an investment package of BRL 952 billion, EUR 420.2 billion equivalent, and target[ed] additional grid-connected electricity generation from renewable sources of 4GW by the end of 2010, 777 MW in 201 and 2 GW in 2012" (ibid.).

British Columbia: This Canadian province enacted a renewable portfolio standard or quota policy (Ren21 2012, 119, table R13). It also introduced a tax of ten dollars per tonne of CO_2 in July (SBS Staff 2012).California: This US state updated the Appliance and Building Efficiency Standards.

China: The several initiatives this year included the 2010 Biomass Electricity Feed-in Tariff, the 2009 Building Integrated Solar PV Program (implemented in 2010), import duty removal on wind and hydro technological equipments, the Interim Feed-in Tariff for four Ningxia solar projects (seen "as a first step in the process of implementing a national feed-in [sic] tariff for solar PV generated electricity," the Interim Measure on the Management of Offshore Wind Farm was intended for either "Chinese-funded companies or Sino-foreign joint ventures" with majority Chinese ownership), and Notice 28 on Import Duty Reduction for Ethanol (to further compliance with WTO standards) (IEA/IRENA 2012).

France: This year marked the enactment of the Renewable Energy Feed-in Tariff: Solar PV, which was then amended in 2011 (ibid.).

India: A carbon tax of fifty rupees per tonne on coal produced or imported to the country was introduced (SBS Staff 2012). This year additionally saw the implementation of the National Solar Mission, projected to run from 2010 to 2022, and promoting both large- and small-scale solar plants (ibid.).

Ireland: A tax on oil and gas was implemented. "It was estimated to add around €43 to filling a 1000 litre oil tank and €41 to the average annual gas bill" (SBS Staff 2012).

Malaysia: A feed-in policy was enacted (Ren21 2012, 118, table R12).

Malta: A feed-in policy was enacted (ibid., 118, table R12).

Puerto Rico: A renewable portfolio standard or quota policy was enacted (ibid., 119, table R13).

South Korea (Republic of Korea): A renewable portfolio standard or quota policy was enacted (ibid., 119, table R13).

United Kingdom: A feed-in policy was enacted (ibid., 118, table R12). This policy offered a feed-in tariff for small-scale renewable energy production and was granted even if households, businesses, or community producers consumed the electricity on-site rather than feeding it back into the grid (IEA/IRENA 2012).

2011   Australia: The Australian Renewable Energy Agency planned to administer 3.2 billion AUD for the RD&D and commercialization of renewable energy "and related technologies" (ibid.). Additionally, a carbon price was planned for introduction in 2012–2013 (ibid.).

California: "On April 12, 2011, Gov. Jerry Brown signed legislation to require one-third of the state's electricity to come from renewable energy by December 31, 2020" (California Energy Commission 2012a). This preempted an earlier renewable portfolio standard formed in 2008 and 2009 by Governor Arnold Schwarzenegger and the California Air Resources Board (California Energy Commission 2012b).

China: A solar PV feed-in tariff was put in place. This year also introduced the twelfth five-year plan for the national economic and social development of the People's Republic of China, which included specific language on renewable energy (IEA/IRENA 2012), including carbon and energy intensity targets.

Denmark: Energy Strategy 2050 was implemented with the goal of achieving "100% independence from fossil fuel in the national energy mix by 2050" (ibid.).

France: The Offshore Wind Tendering Mechanism was established (ibid.).

Germany: The KfW Development Bank launched the Offshore Wind Energy program (ibid.). This same year, Germany released the Sixth Energy Research Program (6.Energieforschunggsprogramm— Forschung für eine Umweltschonende, Zuverlässige und Bezahlbare Energieversorgung), which carried over seven hundred euros million in funding for the years 2011 to 2014, increasing annually (ibid.).

India: The Renewable Energy Certificates System was created (ibid.).

Indonesia: This year saw the establishment of the Tax Incentive for Geothermal Exploration. Regulation 22/PMK.011/2011 reportedly stipulated that "goods imported for the purpose of upstream oil and gas activities or geothermal exploration will be exempted from Value-added Tax for the year 2011" (ibid.).

Israel: A renewable portfolio standard or quota policy was enacted (Ren21 2012, 119, table R13).

Italy: The New Feed-in Premium for Photovoltaic Systems was instituted, superseding the old system implemented in 2005 (ibid.).

Malaysia: The Renewable Energy Feed-in Tariff was established (ibid.).

Netherlands: A feed-in policy was enacted (Ren21 2012, 118, table R12).

Nova Scotia: This Canadian province enacted a feed-in policy (ibid., 118, table R12).

Rhode Island: This US state enacted a feed-in policy (ibid., 118, table R12).

Syria: A feed-in policy was enacted (ibid., 118, table R12).

United Kingdom: The Renewable Heat Incentive consisted of phases, the first of which reportedly included a feed-in tariff scheme for "large emitters in the nondomestic sector" (IEA/IRENA 2012).

United States: CO_2 performance standards for model year 2017– 2025 light-duty automobiles were announced as well as for heavy-duty trucks and buses (EPA 2012b).

Vietnam: Reportedly, "as of August 2011, wind power developers are exempted from import duties on equipment and from corporation tax rates" (IEA/IRENA 2012).

2012 **(Early)**	Australia: Carbon pricing was introduced (ibid.). Austria: Ökostromverordnung (feed-in tariffs) and Ösvo are established (ibid.). Germany: This year saw the passage of the Amendment of the Renewable Energy Sources Act (ibid.). Japan: New feed-in tariffs for solar PV at US$0.53/kWh were implemented (Kraemer 2012). Mexico: The National Energy Plan 2012–2016 was developed (IEA/IRENA 2012). Norway: A renewable portfolio standard or quota policy was enacted (Ren21 2012, 119, table R13). This year also saw the implementation of the Norway-Sweden Green Certification Scheme (IEA/IRENA 2012). Palestinian Territories: A feed-in policy was enacted (Ren21 2012, 118, table R12). Rwanda: A feed-in policy was enacted (ibid., 118, table R12). Spain: The Royal Decree Law 1/2012, Revocation of Public Financial Support for New Electricity Plants from Renewable Energy Sources, Waste, or CHP was passed during this year. Reportedly, "the remuneration pre-assignment registry processes will be temporarily cancelled. The financial support for new installations that produce electricity from renewable energy sources or waste, or for new CHP installations, will also be temporarily abolished. The actual installed capacity from wind power, solar thermal electric and notably that from solar photovoltaic has gone beyond the objectives set up in the Renewable Energy Plan (REP) 2005–2010. Hence, the costs of the financial support for the electricity from renewable energy sources have been significantly higher than had been anticipated" (IEA/IRENA 2012).

Appendix E: Abbreviations

Abbreviations

CAS	Chinese Academy of Sciences
CCS	Carbon capture and storage
CDM	Clean Development Mechanism
CFC	Chlorofluorocarbon
CFL	Compact fluorescent lightbulb
CO_2	Carbon dioxide
EISA	Energy Independence and Security Act
EPA	Environmental Protection Agency
EPAct	Environmental Protection Agency Act
EPO	European Patent Office
ETIS	Energy technology innovation system
EV	Electric vehicle
FGD	Flue-gas desulfurization
GE	General Electric
GW	Gigawatt (one billion watts)
HEV	Hybrid-electric vehicle
IEA	International Energy Agency
IGCC	Integrated gasification combined cycle
IP	Intellectual property
IPC	International patent classification
IRENA	International Renewable Energy Agency
LED	Light-emitting diode
Li-on	Lithium ion (battery)
MOST	Ministry of Science and Technology of China
MW	Megawatt (one million watts)
NDRC	National Development and Reform Commission of China
NGCC	Natural gas combined cycle
NIS	National innovation system
NNSF	National Natural Science Foundation
PHEV	Plug-in electric vehicle

Ppm	Parts per million
PURPA	Public Utility Regulatory Policies Act
PV	Photovoltaic
RMB	Renmenbi (Chinese currency)
R&D	Research and development
RD&D	Research, development, and demonstration
SIPO	State Intellectual Property Office of China
TPRI	Thermal Power Research Institute
TRIPS	Agreement on Trade Related Aspects of Intellectual Property Rights
UNFCCC	UN Framework Convention on Climate Change
WTO	World Trade Organization

Notes

1 Introduction

1. The CDM was established within the Kyoto Protocol in 1997. It permits a country with emission-reduction commitments under the protocol (most industrialized countries) to implement an emission-reduction project in a developing country and get credit for the emission reduction. These emission-reduction credits are known as "offsets." Since 2006, CDM projects have accounted for about three billion tonnes of carbon dioxide equivalent in total, which is equivalent to approximately 10 percent of one year's worth of global carbon dioxide emissions today.

2. John Curtis Perry, Henry Willard Denison Professor at the Fletcher School of Law and Diplomacy, first made this point to me.

3. I thank Graham T. Allison of Harvard University for suggesting that I explore this motive. According to Alan Deardorff (2006, 175), mercantilism is "an economic philosophy of the 16th and 17th centuries that international commerce should primarily serve to increase a country's financial wealth, especially of gold and foreign currency. To that end, exports are viewed as desirable and imports are viewed undesirable unless they lead to even greater exports."

4. The "average" lifetime is the time required for a substance added at a single point in time to decline (at an exponential rate) to 36.8 percent (1/e) of the original amount. In concept, it is similar to the half-life, which is the time required for a substance to decrease to one-half of its initial value. To Nike's great credit, it phased out the use of sulfur hexafluoride in its Nike Air shoes voluntarily in order to reduce Nike's contribution to the climate change problem. For more information, please see: http://americancarbonregistry.org.

5. CO_2 does not follow a simple exponential decline because multiple processes remove it at different rates. The reported hundred-year lifetime is the time it takes for a pulse of CO_2 added to the atmosphere to decline to 36.8 percent of an original concentration. But the removal rate slows over time, and there is a long "tail" of concentration that gradually decreases for many hundreds of years.

6. Technologies considered "clean" by this study include: all biomass, geothermal, and wind-generation projects of more than 1 megawatt (MW), all hydro

projects of between 0.5 and 50 megawatts, all solar projects of more than 0.3 megawatts, all marine energy projects, and all biofuel projects with a capacity of 1 million liters or more per year.

7. The data for this export comparison assessment were sourced from the United Nations' COMTRADE database, the "largest depository of international trade data," which includes trade figures from over 170 countries coded according to standardized formats. The COMTRADE queries were executed through the World Bank's WITS interface.

8. I first heard a version of this formulation from Professor John P. Holdren during graduate school in a class at the Harvard Kennedy School in the late 1990s.

9. Even these can have liabilities. A good example would be the compact fluorescent lightbulb (CFL), which can reduce electricity use from lighting by upward of 75 percent. CFLs contain a small amount of mercury, which is toxic to humans and animals if not disposed of properly. The US Environmental Protection Agency (EPA) estimates that CFL usage results in considerably less mercury emissions compared with incandescent lightbulbs, however, because of the mercury emissions associated with electricity production. According to the EPA, if every home in the United States replaced just one incandescent lightbulb with an ENERGY STAR qualified CFL, enough energy would be saved to light more than three million homes. Or the savings would prevent the release of greenhouse gas emissions equal to that of about eight hundred thousand cars (US EPA 2010).

10. This "conventional" versus "unconventional" framing is attributed to Lema and Lema 2010.

11. For a report from this workshop, see http://fletcher.tufts.edu/CIERP.

2   Into the Dragon's Den

1. This section and the next one draw heavily from Gallagher and Lewis 2012.

2. Welfare loss is a combination of losses in consumption and leisure time.

3. Defined as those enterprises from nine industrial sectors that consume 180,000 tons of coal equivalent (0.005 exajoule) or higher annually.

4. This is the strategy that was also used by China when it shifted away from ozone-depleting CFCs, which led to it becoming the world's leading producer of compressors, refrigerators, and window air conditioners (Moomaw 2012).

3   Four Telling Tales

1. Interview 66, 2010. The interviewee was assuming that China's electricity sector would remain dependent on fossil fuels for decades to come, but would shift to IGCC technology coupled with CCS as well as natural gas power generation.

2. Interview 66, 2010.

3. Interview 48, 2010.

4. Interview 66, 2010.

5. Interview 7, 2010.

6. Interview 76, 2010.

7. Ibid.

8. Interview 41, 2012.

9. Interviews 41, 2012, and 51, 2011.

10. Interview 91, 2010.

11. Interviews 65 and 71, 2010.

12. Interview 21, 2010.

13. Interview 71, 2010.

14. Interview 65, 2010.

15. Interview 33, 2010.

16. Interview 77, 2010.

17. Interview 61, 2012.

18. Interview 18, 2010.

19. Interviews 21 and 56, 2010.

20. Interviews 18, 65, and 77, 2010.

21. Interview 56, 2010.

22. Interview 65, 2010.

23. Ibid.

24. Interview 30, 2012.

25. Ibid.

26. Interview 59, 2010.

27. Interviews 30, 2012, and 43, 2010.

28. Chinese firms pointed out that SolarWorld received tax breaks and subsidies worth $43 million in Oregon in 2007, and had received 136 million euros worth of government support in the European Union since 2003 (Osborne 2011). The Coalition for Affordable Solar Energy compiled a list of publicly available information about SolarWorld's government support and then issued a report detailing subsidies provided to SolarWorld alone at more than $100 million, ranging from the US clean energy manufacturing tax credit of $82.2 million, a 100 percent local property tax abatement for up to five years, an $11 million business energy tax credit in Oregon, and more (Choudhury 2012).

29. Interview 30, 2012.

30. Ibid.

31. Interview 57, 2010.

32. Interview 59, 2010.

33. Interview 58, 2011.

34. Interviews 58 and 59, 2011.

35. Interview 57, 2010.

36. Interview 59, 2010.

37. Interview 30, 2012.

38. Interview 59, 2010.

39. Interview 66, 2010.

40. Interviews 35 and 37, 2010.

41. Interview 48, 2010.

42. Interview 41, 2010.

43. Interviews 34, 37, and 66, 2010.

44. Interview 66, 2010.

45. Interviews 37, 2010, and 41, 2012.

46. Interview 34, 2010.

4   The Essential Role of Policy

1. The one exception here is that few countries need to truly achieve energy "independence" because oil markets are global and natural gas markets are becoming so. If a country needs to buy oil, it can do so on the global oil market. One country's boycott of another should not affect a consuming country from being able to buy oil from others. Indeed, achieving true independence could come at great cost because the global oil market is likely to provide much lower prices than can be obtained domestically. This is less true for natural gas, which cannot be traded as easily when the gas has to move by pipeline.

2. Interview 61, 2011.

3. Interview 31, 2012.

4. Interviews 37, 2010, and 41, 2012.

5. Interview 79, 2011.

6. In an excellent case study of Goldwind, Joanna Lewis (2013) also documents the large amount of finance made available to the firm for R&D and overseas ventures, including a US$6 billion credit line from the China Development Bank to expand its business in the United States.

7. Interview 58, 2011.

8. Interviews 37, 2010, and 41, 2012.

9. Interview 22, 2011.

10. This point was made most forcefully in interview 31, 2012.

11. Interview 62, 2011.

12. Many interviewees contributed to this list, but a few were the most specific and forceful: interview 41, 2012; interview 47, 2012; interview 59, 2010; interview 61, 2011. Some of these insights are attributed to Global Energy Assessment 2012, and using different wording, IPCC 2000.

13. While it was beyond the scope of this project, it would be interesting and useful to develop correlations technology by technology with individual policy initiatives, if they could be differentiated by technology.

14. These figures were calculated from data obtained through Comtrade (2012) and the World Development Indicators (World Bank 2012).

5   No Risk, No Reward

1. Interview 65, 2010.

2. Wanxiang America Corporation acquired A123 out of bankruptcy on January 29, 2013. All the research for this book was conducted before this event occurred.

3. The term n-1 or n-2 came from one of the interviews, and it means previous generations of technologies—in other words, the current generation minus 1.

4. Alford (1995, 123) wrote, "If it is true that serious protection for foreign intellectual property in the PRC must await further development of Chinese-generated intellectual property of commercial importance, it follows that a PRC willing to accord American holders of intellectual property more of the rights they now seek will likely have many more enterprises that are technologically sophisticated and increasingly commercially competitive internationally. In short, the conditions that breed protection for intellectual property are also those that breed competition with regard to intellectual property."

5. Li's study is overall and not exclusively focused on energy patents.

6. Interviews 9 and 61, 2011.

7. Interviews 9 and 61, 2011; interviews 31 and 30, 2012; interview 5, 2011.

8. See http://ipr.court.gov.cn/ (accessed June 30, 2013).

9. It would be nice to be able to search a similar database of arbitration cases, but none exists that the author is aware of.

10. As of the writing of this book, the two parties would not agree to be interviewed, and the outcomes of this situation were not yet apparent, so this significant case could not be examined in detail.

11. See http://www.wto.org/english/tratop_e/dispu_e/dispu_by_country_e.htm (accessed June 30, 2013).

12. Zhang Fang of Tsinghua University suggested this hypothesis.

13. The author gratefully acknowledges the research assistance of Amos Irwin for this section.

14. One, non-peer-reviewed study asserted that SIPO patent examiners are well regarded internationally (Harvey 2011).

15. For information on which IPC codes were used and how they were selected, see appendix C.

16. In other words, one cannot protect a "method" unless it is embodied in a product. SIPO's 2010 Guidelines for Examination state that a utility model patent can be granted only for "products." "The products herein shall be objects manufactured by industrial methods, having definite shape and structure, and

occupying a certain space. Utility model patents do not cover manufacturing processes, methods of use, methods of communication, processing methods, computer programs, or the method of applying a product to a specific purpose" (SIPO 2010). In other words, most clean and efficient energy technologies would not qualify for a utility model patent in China.

17. The cases discussed in these interviews were not related to the energy industry. At the time of this writing, the only clean-energy-related court case that the author is aware of is the ongoing dispute between American Superconductor and Sinovel over wind turbine technology.

18. Interviews 48, 66, and 76, 2010.

6   Competing against Incumbents

1. "New combinations" was Schumpeter's term for technological innovations.

2. In a cap-and-trade program, emissions are capped at a certain level, and permits to emit the pollutant are distributed or auctioned. Permit holders can then trade credits through an exchange.

3. See http://www.eia.gov/oiaf/beck_plantcosts (accessed July 1, 2013).

4. This essential idea, if not the exact formulation, is attributed to John P. Holdren.

5. Interviews 37 and 34, 2010.

6. Zhang Fang, a doctoral candidate at Tsinghua University, conceived of the useful metaphor of the snowball during a discussion I had with her about access to financing in China.

7. Interview 41, 2012.

8. Interviews 43 and 65, 2010.

9. Interview 43, 2010.

10. Interview 65, 2010.

11. Ibid.

12. Interview 71, 2010.

13. Interviews 65 and 71, 2010.

14. Interview 34, 2010.

15. Interview 37, 2010.

16. Interview 30, 2012.

17. Interview 66, 2010.

7   The Global Diffusion of Cleaner Energy Technologies

1. These figures were calculated from data obtained through Comtrade (2012) and the World Development Indicators (World Bank 2012).

2. The data on batteries only became available in 2010.

Appendix D

1. There are a few key reasons why the PURPA policy of 1978 is considered to be the first feed-in tariff policy: it required utilities to buy electricity generated from qualifying RE facilities at preestablished rates (Lipp 2007); it was production based, awarding per kWh payments for all the electricity generated from the facility, rather than simply the surplus; in certain cases, the payment levels were differentiated by technology type; and finally, in some jurisdictions such as California, it was implemented using long-term contracts for electricity sales (Hirsh 1999). PURPA contracts slowed down with electricity restructuring in the late 1990s, and few are signed today because utilities have turned to other means of procuring electricity from renewable energy sources (Couture et al. 2010, 9n15). States, not the DOE (2012c), implement PURPA. In 2005, new federal standards on net metering, fuel sources, fossil fuel generation efficiency, time-based metering and communications, and interconnection were added to PURPA section 111(d). The 2007 Energy Independence and Security Act (EISA) added four new federal standards to section 111(d), including (verbatim) integrated resource planning, rate design modifications to promote energy-efficiency investments, consideration of smart grid investments, and smart grid information (ibid.). The impact of these Energy Policy Act (EPAct) 2005 and EISA 2007 changes to the 1978 PURPA law is that state electricity regulators (i.e., state public utility commissions) "must consider," for their regulated electric utilities (usually but not always only investor-owned utilities), whether to adopt verbatim all these standards as requirements on those electric utilities. By must consider, PURPA as amended says that states must start regulatory proceedings by a specified deadline, and then make a yes or no decision by another specified date on whether to actually adopt that standard verbatim as a requirement on its state-jurisdictional utilities (ibid.).

2. In December 1990, the first national feed-in tariff legislation in Europe was adopted in Germany's Electricity Feed-in Law (Stromeinspeisungsgesetz or StrEG). As of January 1, 1991, utilities in Germany were required by law to buy electricity from nonutility renewable energy generators at a fixed percentage of the retail electricity price (Rickerson, Sawin, and Grace 2007). The StrEG included a purchase obligation for this electricity, and the percentage ranged from 65 to 90 percent depending on the technology type and project size. A project size cap of five MW was also imposed on hydropower, landfill gas, sewage gas, and biomass facilities. Denmark and Spain followed suit with similar provisions in 1992 and 1997, respectively (Couture et al. 2010, 9). The policy approach was later abandoned and another adopted (Couture et al. 2010, 20). Certain municipal utilities also began (although it is unclear if this was in the same year) to offer feed-in tariff prices "based on the actual costs of RE generation" instead of avoided cost pricing (ibid., 9). A new feed-in tariff was adopted in 2000, and then revised in 2004 and 2008 (Grace, Rickerson, and Corfee 2009, 17).

References

Aanesen, Krister, Stefan Heck, and Dickon Pinner. 2012. *Solar Power: Darkest before Dawn.* New York: McKinsey and Co.

Abdel-Latif, Ahmed. 2013. Ways to Promote Enabling Environments and to Address Barriers to Technology Development and Transfer. ICTSD Paper. http://www.ictsd.org/downloads/2012/09/ways-to-promote-enabling-environments.pdf (accessed July 9, 2013).

Abernathy, William, and James M. Utterback. 1978. Patterns of Innovation in Technology. *Technology Review* 80 (7): 40–47.

Ackerman, Frank, and Lisa Heinzerling. 2004. *Priceless: On Knowing the Price of Everything and the Value of Nothing.* New York: New Press.

Ackerman, Frank, Elizabeth A. Stanton, Chris Hope, and Stephane Alberth. 2009. Did the Stern Review Underestimate US and Global Climate Damages? *Energy Policy* 37 (7): 2717–2721.

Alford, William P. 1995. *To Steal a Book is an Elegant Offense: Intellectual Property Law in Chinese Civilization.* Stanford, CA: Stanford University Press.

Allison, Graham T. 2004. *Nuclear Terrorism—The Ultimate Preventable Catastrophe.* New York: Henry Holt.

Amsden, Alice H. 2001. *The Rise of "The Rest": Challenges to the West from Late-Industrializing Economies.* New York: Oxford University Press.

Andersen, Mikael S. 2012. Europe's Experience with Carbon-Energy Taxation. *Surveys and Perspectives Integrating Environment and Society* 3 (2): 1–12.

Anonymous. 2011a. Guangdong: Guangzhou IP Office Reviewed 3,000 Applications for Patent Subsidies over the Internet. Ministry of Commerce, People's Republic of China. http://www.chinaipr.gov.cn (accessed February 17, 2012).

Anonymous. 2011b. Hunan: Applications for Patent Subsidies to End on August 30. Ministry of Commerce, People's Republic of China. http://www.chinaipr.gov.cn (accessed February 17, 2012).

Anthoff, David, Richard Tol, and Gary W. Yohe. 2009. Risk Aversion, Time Preference, and the Social Cost of Carbon. *Environmental Research Letters* 4 (2): 1–7.

Argote, Linda, and Dennis Epple. 1990. Learning Curves in Manufacturing. *Science* 247 (4945): 920–924.

Arthur, W. Brian. 2009. *The Nature of Technology*. New York: Free Press.

Auer, Matthew. 1993. Transferring Gas-Turbine Technologies to the Former Soviet Union: Opportunities and Problems. *Journal of Technology Transfer* 18 (1–2): 55–61.

Barton, John H. 2007. Intellectual Property and Access to Clean Energy Technologies in Developing Countries. ICTSD Trade and Sustainable Energy Series, issue 2. http://ictsd.org/i/publications/3354.

Basberg, Bjorn L. 1987. Patents and the Measurement of Technological Change: A Survey of the Literature. *Research Policy* 16 (2–4): 131–141.

Beér, Janos. 2009. Higher Efficiency Power Generation Reduces Emissions. National Coal Council Issue Paper, MIT, Cambridge, MA. http://web.mit.edu/mitei/docs/reports/beer-emissions.pdf (accessed June 13, 2012).

Bell, Martin. 2012. International Technology Transfer, Innovation Capabilities, and Sustainable Directions of Development. In *Low Carbon Technology Transfer: From Rhetoric to Reality*, ed. David G. Ockwell and Alexandra Mallett. Abingdon, UK: Earthscan.

Bell, Martin, and Michael Albu. 1999. Knowledge Systems and Technological Dynamism in Industrial Clusters in Developing Countries. *World Development* 27 (9): 1715–1734.

Bell, Martin, and Keith Pavitt. 1993. Technological Accumulation and Industrial Growth: Contrasts between Developed and Developing Countries. *Industrial and Corporate Change* 2 (2): 157–210.

Berkhout, Frans, Adrian Smith, and Andy Stirling. 2004. Socio-Technological Regimes and Transition Contexts. In *System Innovation and the Transition to Sustainability: Theory, Evidence, and Policy*, ed. Boelie Elzen, Frank W. Geels, and Kenneth Green. Cheltenham, UK: Edward Elgar.

Birner, Sabrina, and Eric Martinot. 2005. Market Transformation for Energy-Efficient Products: Lessons from Programs in Developing Countries. *Energy Policy* 33 (18): 1765–1779.

Bloomberg News. 2012. China to Give Stimulus for Development of Electric Vehicles. Bloomberg, April 18.

Bohi, Douglas R., and Michael Toman. 1996. *The Economics of Energy Security*. Norwell, MA: Kluwer Academic Publishers.

Bolinger, Mark, and Ryan Wiser. 2012. Understanding Wind Turbine Price Trends in the U.S. over the Past Decade. *Energy Policy* 42 (3): 628–641.

BP (British Petroleum). 2011. Statistical Review of World Energy. London: British Petroleum.

BP. 2012. Statistical Review of World Energy. London: British Petroleum.

Branstetter, Lee, Rayond Fisman, and Fritz Foley. 2006. Do Stronger Intellectual Property Rights Increase International Technology Transfer? Empirical Evidence from U.S. Firm-Level Panel Data. *Quarterly Journal of Economics* 121 (February): 321–349.

Branstetter, Lee, Raymond Fisman, C. Fritz Foley, and Kamal Saggi. 2007. Intellectual Property Rights, Imitation, and Foreign Direct Investment: Theory and Evidence. NBER Working Paper, 13033. Cambridge, MA: National Bureau of Economic Research.

Brewer, Thomas. 2008. Climate change technology transfer: a new paradigm and policy agenda. *Climate Policy* 8 (5): 516–526.

Brooks, Harvey. 1995. What We Know and Do Not Know about Technology Transfer: Linking Knowledge to Action. In Marshalling Technology for Development: Proceedings of a Symposium, ed. National Academy of Sciences. Washington, DC: National Academy Press.

Brown, Marilyn A. 2001. Market Failures and Barriers as a Basis for Clean Energy Policies. *Energy Policy* 29 (14): 197–207.

Brown, Marilyn A., and Benjamin K. Sovacool. 2011. Barriers to the Diffusion of Climate Friendly Technologies. *International Journal of Technology Transfer and Commercialisation* 10 (1): 43–62.

Buen, Jorund. 2006. Danish and Norwegian Wind Industry: The Relationship between Policy Instruments, Innovation, and Diffusion. *Energy Policy* 34 (18): 3887–3897.

Bunn, Matthew. 2010. Securing the Bomb: Securing all Nuclear Materials in Four Years. Washington, DC: Nuclear Threat Initiative. http://www.nti.org/securingthebomb.

Bureau of Labor Statistics. 2011. Automotive Industry: Employment, Earnings, and Hours. Washington, DC: US Government Printing Office. http://www.bls.gov/iag/tgs/iagauto.htm#emp_national (accessed January 2012).

Butler, Lucy, and Karsten Neuhoff. 2008. Comparison of Feed-In Tariff, Quota, and Auction Mechanisms to Support Wind Power Development. *Renewable Energy* 33 (8): 1854–1867.

Cai, Wenjia, Can Wang, Wenling Liu, Ziwei Mao, Huichao Yu, and Jining Chen. 2009. Sectorial Analysis for International Technology Development and Transfer: Cases of Coal-Fired Power Generation, Cement, and Aluminium in China. *Energy Policy* 37 (6): 2283–2291.

California Energy Commission. 2012a. Renewable Energy Programs. http://www.energy.ca.gov/renewables/renewable_links.html (accessed July 27, 2012).

California Energy Commission. 2012b. Renewables Portfolio Standards (RPS) Proceeding Docket # 11-RPS-01 and 03-RPS-1078. http://www.energy.ca.gov/portfolio/index.html (accessed July 27, 2012).

California Public Utilities Commission. 2012. § 399.20 Feed-in Tariff (FIT) Program for the Purchase of Eligible Small Renewable Generation. Last modified August 31, 2012. http://www.cpuc.ca.gov/PUC/energy/Renewables/hot/feedin-tariffs.htm (accessed September 13, 2012).

Cao, Cong, Richard P. Suttmeier, and Denis Simon. 2006. China's 15-Year Science and Technology Plan. *Physics Today* 59 (12): 38–43.

Carlson, Ann E., and Robert W. Fri. 2013. Designing a Durable Energy Policy. *Daedalus* 142 (1): 119–128.

Carlsson, Bo, and Rikard Stankiewicz. 1995. On the Nature, Function, and Composition of Technological Systems. In *Technological Systems and Economic Performance: The Case of Factory Automation*, ed. Bo Carlsson. Berlin: Kluwer Academic Publishers.

Casey, Joseph, and Katherine Koleski. 2011. Backgrounder: China's 12th Five-Year Plan. Washington, DC: US-China Economic and Security Review Commission.

Casey, Zoe. 2013. Is German Offshore Wind under Threat? European Wind Energy Association. http://www.ewea.org/blog/2013/04/is-german-offshore-wind -under-threat/ (accessed May 15, 2013).

Chang, Ha-Joon. 2002. *Kicking Away the Ladder: Development Strategy in Global Perspective*. London: Anthem Press.

Chang, Jackie. 2011. Half of China Solar Firms Halt Production, Says Report. Digitimes, Taipai. http://www.digitimes.com/news/a20111209VL200.html?mod =3&q=SOLAR (accessed April 27, 2012).

Chen, Aizhu. 2012. Lagging New Capacity to Strain China's Power Supply This Year. Reuters, January 16, 2012.

Chen, Wenying, and Ruina Xu. 2010. Clean Coal Technology Development in China. *Energy Policy* 38 (5): 2123–2130.

China Daily. 2012a. China Plans African Ventures. *China Daily*, June 8. http:// www.chinadaily.com.cn/business/2011-06/08/content_12657345.htm (accessed June 29, 2013).

China Daily. 2012b. China Waives Sales Tax on Locally Made EVs, Fuel Cell Cars. *China Daily*, January 9.

Choudhury, Nilima. 2012. CASE Accuses CASM Members of Hypocrisy. PV-Tech, March 20.

CIA (Central Intelligence Agency). 2012. CIA World Factbook in U.S. Central Intelligence Agency. Washington, DC: CIA. https://www.cia.gov/library/ publications/the-world-factbook/rankorder/2172rank.html (accessed May 17, 2012).

Clark, Norman, and Calestous Juma. 1987. *Long-Run Economics: An Evolutionary Approach to Economic Change*. London: Pinter Publishers.

Cohen, Marc. 2010. Presentation at Microsoft Beijing Office, July 11.

Cohen, Wesley M., Richard R. Nelson, and John P. Walsh. 2000. Protecting Their Intellectual Assets: Appropriability Conditions and Why U.S. Manufacturing Firms Patent (or Not). NBER Working Paper. Cambridge, MA: National Bureau of Economic Research.

Comtrade. 2012. United Nations Commodity Trade Statistics Database, Statistics Division. http://comtrade.un.org.

Copenhagen Economics. 2009. Are IPRs a Barrier to the Transfer of Climate Change Technology? http://trade.ec.europa.eu/doclib/docs/2009/february/tradoc _142371.pdf.

Corvaglia, Maria Anna. 2010. South-South Technology Transfer Addressing Climate Change. http://www.nccr-climate.unibe.ch/conferences/climate_economics _law/papers/Corvaglia_MariaAnna.pdf (accessed July 9, 2013).

Couture, Toby D., Karlynn Cory, Claire Kreycik, and Emily Williams. 2010. A Policymaker's Guide to Feed-in Tariff Policy Design. Technical Report NREL/ TP-6A2–44849. July. Golden, CO: National Renewable Energy Laboratory. http://www.osti.gov/bridge (accessed July 8, 2013).

Cozzi, Paolo. 2012. Assessing Reverse Auctions as a Policy Tool for Renewable Energy Deployment. Center for International Environment and Resource Policy Discussion Paper 007. The Fletcher School, Tufts University, Medford, MA.

Dagvadorj, D. 2006. Technology Transfer and Barriers in Mongolia. https:// www.google.com/url?q=http://unfccc.int/files/documentation/workshops _documentation/application/pdf/moncp.pdf&sa=U&ei=cQwAUq_XOYbLqQG btYG4Dw&ved=0CAcQFjAA&client=internal-uds-cse&usg=AFQjCNH7FAnc qAZ1K3VUgv8MIg0eNRUMhw.

Davidson, Ogunlade R., and Frank Teng-Zeng. 2003. Transfer of Emerging Energy Technologies. http://www.techmonitor.net/tm/images/9/96/03nov_dec _sf1.pdf (accessed July 9, 2013).

de la Tour, Arnaud, Matthieu Glachant, and Yann Ménière. 2011. Innovation and International Technology Transfer: The Case of the Chinese Photovoltaic Industry. *Energy Policy* 39 (2): 761–770.

Deardorff, Alan. 2006. *Terms of Trade: Glossary of International Economics.* Paris: Lavoisier Librairie.

Dechezlepretre, Antoine, Matthieu Glachant, Ivan Hascic, Nick Johnstone, and Yann Ménière. 2011. Invention and Transfer of Climate Change–Mitigation Technologies: A Global Analysis. *Review of Environmental Economics and Policy* 5 (1): 109–130.

del Rio, Pablo, and Miguel A. Gual. 2007. An Integrated Assessment of the Feed-in Tariff System in Spain. *Energy Policy* 35 (2): 994–1012. http://dx .doi.org.ezproxy.library.tufts.edu/10.1016/j.enpol.2006.01.014 (accessed July 16, 2012).

Dittmar, Cynthia. 2008. Barriers and Drivers for the Deployment of Renewable Energy Technology in Developing Countries Case Study: Solar Water Pumps in Ghana. http://www.grin.com/en/e-book/131922/barriers-and-drivers-for-the -deployment-of-renewable-energy-technology (accessed July 9, 2013).

Dixon, Robert, Richard M. Scheer, and Gareth T. Williams. 2011. Sustainable Energy Investments: Contributions of the Global Environment Facility. *Mitigation and Adaptation Strategies for Global Change* 16 (1): 83–102.

DOE (Department of Energy). 2010. *Critical Materials Strategy.* Washington, DC: US Department of Energy.

DOE. 2012a. A Banner Year for the U.S. Wind Industry. Washington, DC: US Department of Energy. http://energy.gov/articles/banner-year-us-wind-industry (accessed September 9, 2012).

DOE. 2012b. DOE Loan Programs Office. Washington, DC: US Department of Energy. https://lpo.energy.gov/ (accessed September 30, 2012).

DOE. 2012c. Public Utility Regulatory Policies Act of 1978 (PURPA). http://energy.gov/oe/services/electricity-policy-coordination-and-implementation/other -regulatory-efforts/public (accessed July 19, 2012).

DOE. 2012d. Sunshot Initiative Mission, Vision, Goals. Washington, DC: US Department of Energy. http://www1.eere.energy.gov/solar/sunshot/mission _vision_goals.html (accessed August 20, 2012).

Dosi, Giovanni. 1982. Technological Paradigms and Technological Trajectories. *Research Policy* 11:147–162.

Dosi, Giovanni, Christopher Freeman, Richard R. Nelson, Gerald Silverberg, and Luc L. Soete, eds. 1988. *Technology and Economic Theory*. London: Pinter.

Earley, Robert, Liping Kang, Feng An, and Lucia Green-Weiskel. 2011. Electric Vehicles in the Context of Sustainable Development in China. New York: UN Department of Economic and Social Affairs, Commission on Sustainable Development, 9, CSD19/2011/BP9.

Edquist, Charles, ed. 1997. *Systems of Innovation: Technologies, Institutions, and Organizations*. London: Pinter.

EIA (Energy Information Administration). 2011. *Electric Power Annual, 2010*. Washington, DC: US Department of Energy, Energy Information Administration.

EIA. 2012a. Annual Energy Outlook. DOE/EIA-0388. Washington, DC: US Department of Energy.

EIA. 2012b. Electricity Data Browser. Washington, DC: US Department of Energy. http://www.eia.gov/beta/enerdat/ (accessed August 21, 2012).]

EIA. 2012c. Most States Have Renewable Portfolio Standards. Washington, DC: US Department of Energy. http://www.eia.gov/todayinenergy/detail.cfm?id=4850 (accessed April 12, 2012).

Energy Efficiency and Renewable Energy. 2012. American Recovery and Rein-vestment Act. Washington, DC: US Department of Energy. http://www1 .eere.energy.gov/recovery/buy_american_provision.html (accessed September 26, 2012).

Environmental Administration. 2012. Environmentally Related Energy Taxation in Finland. Last modified October 2, 2012. http://www.environment.fi/default .asp?contentid=147208&lan=en (accessed July 27, 2012).

EPA. 2012a. History of the Clean Air Act. Last modified February 17, 2012. http://epa.gov/oar/caa/caa_history.html (accessed September 13, 2012).

EPA. 2012b. Renewable Fuel Standard (RFS). Last modified March 26, 2012. http://www.epa.gov/oms/fuels/renewablefuels/index.htm (accessed July 8, 2012).

Ernst, Dieter, and Linsu Kim. 2002. Global Production Networks, Knowledge Diffusion, and Local Capability Formation. *Research Policy* 31 (8–9): 1417–1429.

Europa. 2012. European Union. http://www.europa.eu/legislation_summaries/energy/renewable_energy/I27035_en.html (accessed September 19, 2012).

Federal Ministry for the Environment, Nature Conservation, and Nuclear Safety. 2011. Renewably Employed. Berlin: Federal Ministry for the Environment, Nature Conservation, and Nuclear Safety.

Ferioli, Francesco, Koen Schoots, and Bob C. C. van der Zwaan. 2009. Use and Limitations of Learning Curves for Energy Technology Policy: A Component-Learning Hypothesis. *Energy Policy* 37 (7): 2525–2535.

Financial Times. 2012. Estimated Shale Gas in Relation to Conventional Gas Reserves. *Financial Times*, April 23.

First Solar. 2011. First Solar to Build Solar Module Factory in Mesa, Arizona. Press release. http://www.investor.firstsolar.com/releasedetail.cfm?releaseid=573730.

Forsyth, Tim. 2003. Climate Change Investment and Technology Transfer in Southeast Asia. *Climate Change and East Asia: The Politics of Global Warming in China and East Asia*, ed. Paul G. Harris, 237–257. London: Routledge.

Freeman, Christopher. 1982. Technological Infrastructure and International Competitiveness. Draft paper submitted to the Organization for Economic Cooperation and Development, Ad Hoc Group on Science, Technology, and Competitiveness.

Freeman, Christopher. 1987. *Technology Policy and Economic Performance: Lessons from Japan*. London: Pinter.

Fu, Xiaolan, and Jing Zhang. 2011. Technology Transfer, Indigenous Innovation, and Leapfrogging in Green Technology: The Solar-PV Industry in China and India. *Journal of Chinese Economic and Business Studies* 9 (4): 329–347.

FuturePolicy. 2011. Indonesia. Last modified November 2, 2011. http://www.futurepolicy.org/2613.html (accessed September 12, 2012).

Gallagher, Kelly Sims. 2006a. *China Shifts Gears: Automakers, Oil, Pollution, and Development*. Cambridge, MA: MIT Press.

Gallagher, Kelly Sims. 2006b. Limits to Leapfrogging in Energy Technologies? Evidence from the Chinese Automobile Industry. *Energy Policy* 34 (4): 383–394.

Gallagher, Kelly Sims. 2006c. Roundtable on Barriers and Incentives for Hybrid Vehicles in China. Report on a roundtable jointly organized by the Energy Technology Innovation Project of the Belfer Center for Science and International Affairs at Harvard University's John F. Kennedy School of Government, Cambridge, MA, and the China Automotive Technology and Research Center, Beijing.

Gallagher, Kelly Sims. 2013. Why and How Governments Support Renewable Energy. *Daedalus* 142 (1): 1–19.

Gallagher, Kelly Sims, Laura Diaz Anadon, Ruud Kempener, and Charlie Wilson. 2011. Trends in Investments in Global Energy Research, Development, and Demonstration. *Wiley Interdisciplinary Reviews: Climate Change* 2 (3): 373–396.

Gallagher, Kelly Sims, Arnulf Grübler, Laura Kuhl, Gregory F. Nemet, and Charlie Wilson. 2012. The Energy Technology Innovation System. *Annual Review of Environment and Resources* 37.

Gallagher, Kelly Sims, John P. Holdren, and Ambuj D. Sagar. 2006. Energy-Technology Innovation. *Annual Review of Environment and Resources* 31 (November): 193–237.

Gallagher, Kelly Sims, and Joanna I. Lewis. 2012. China's Quest for a Green Economy. In *Environmental Policy: New Directions for the Twenty-First Century*, ed. Norman Vig and Michael Kraft. 8th ed. Los Angeles: Sage/CQ Press.

Gallagher, Kelly Sims, J. R. Siegel, and Aaron Strong. 2011. Harnessing Energy: Technology Innovation in Developing Countries to Achieve Sustainable Prosperity. Background paper prepared for the World Economic and Social Survey, Department of Economic and Social Affairs of the UN Secretariat, New York.

Garud, Raghu. 1997. On the Distinction between Know-how, Know-why, and Know-what. *Advances in Strategic Management* 14:81–101.

Gaston, Christian. 2009. State Tax Break Aids SolarWorld, Wal-mart. *Portland Tribune*, October 30.

Gboney, William. 2009. Promoting Tech Transfer and Deployment for Renewable Energy and Energy Efficiency in Ghana. http://www.eprg.group.cam.ac.uk/wp-content/uploads/2009/09/isda_ghana-country-study_september-2009-report_2.pdf (accessed July 9, 2013).

Geels, Frank W., and Johan Schot. 2007. Typology of Sociotechnical Transition Pathways. *Research Policy* 36 (3): 399–417.

Global Energy Assessment. 2012. *Global Energy Assessment*. Cambridge: Cambridge University Press.

Global Environment Facility. 2010. Transfer of Environmentally Sound Technologies: Case Studies from the GEF Climate Change Portfolio. http://www.thegef.org/gef/pubs/tech-transfer-case-studies-2010 (accessed July 9, 2013).

Goldemberg, José. 1998. Leapfrog Energy Technologies. *Energy Policy* 2 (10): 729–741.

Goldman, Daniel P., John J. McKenna, and Lawrence M. Murphy. 2005. *Financing Projects That Use Clean-Energy Technologies: An Overview of Barriers and Opportunities*. Golden, CO: National Renewable Energy Laboratory.

Goossens, Ehren. 2012. AMSC Taking Sinovel Infringement Suit to China's Supreme Court. Bloomberg, April 9.

Government of Brazil. 2010. New Decree Details Brazil's National Policy on Climate Change. http://www.brasil.gov.br/news/history/2010/12/10/new-decree-details-brazil2019s-national-policy-on-climate-change/newsitem_view?set_language=en (accessed September 19, 2012).

Grace, Robert, William Rickerson, and Karin Corfee (KEMA). 2009. *California Feed-In Tariff Design and Policy Options.* California Energy Commission. Publication Number: CEC-300-2008-009F.

Grau, Thilo, Molin Huo, and Karsten Neuhoff. 2011. *Survey of Photovoltaic Industry and Policy in Germany and China.* Berlin: Deutsches Institute für Wirtschaftsforschung.

Greene, David L. 2010. Measuring Energy Security: Can the United States Achieve Oil Independence? *Energy Policy* 38 (4): 1614–1621.

Griliches, Zvi. 1990. Patent Statistics as Economic Indicators: A Survey. *Journal of Economic Literature* 28 (4): 1661–1707.

Grübler, Arnulf. 1998. *Technology and Global Change.* Cambridge: Cambridge University Press.

Grübler, Arnulf. 2010. The Costs of the French Nuclear Scale-Up: A Case of Negative Learning by Doing. *Energy Policy* 38 (9): 5174–5188.

Grübler, Arnulf, F. Aguayo, Kelly Sims Gallagher, Marko P. Hekkert, Kejun Jiang, L. Mytelka, Lena Neij, Gregory F. Nemet, and Charlie Wilson. 2012. Policies for the Energy Technology Innovation System. In *Global Energy Assessment.* Cambridge: Cambridge University Press.

Grübler, Arnulf, Nebojsa Nakicenovic, and David G. Victor. 1999. Dynamics of Energy Technologies and Global Change. *Energy Policy* 27 (5): 247–280.

Hamilton, Kirsty. 2009. *Unlocking Finance for Clean Energy: The Need for "Investment Grade" Policy.* London: Chatham House.

Hammer, Alexander, and Katherine Linton. 2011. China: Effects of Intellectual Property Infringement and Indigenous Innovation Policies on the U.S. Investigation No. 332-519. USITC Publication 4226. Washington, DC: US International Trade Commission.

Han, Tianyang. 2011. Taxes Slashed to Cut Emissions. *China Daily*, June 20.

Hansen, Ulrich E. 2011. An Empirical Case Study of the Transfer of GHG Mitigation Technologies from Annex I Countries to Malaysia under the Kyoto Protocol's Clean Development Mechanism (CDM). *International Journal of Technology and Commercialization* 10 (1): 1–20.

Harvey, Ian. 2011. *Intellectual Property: China in the Global Economy: Myth and Reality.* London: International IP Strategists Association.

Harwitt, Eric. 1995. *China's Automobile Industry: Policies, Problems, and Prospects.* New York: M. E. Sharpe.

Haum, Rüdiger. 2012. Project-Based Market Transformation in Developing Countries and International Technology Transfer: The Case of the Global Environment Facility and Solar PV. *Low Carbon Technology Transfer: From Rhetoric to Reality*, ed. David G. Ockwell and A. Mallett. Abingdon, UK: Earthscan.

Hekkert, Marko P., and Simona O. Negro. 2009. Functions of Innovation Systems as a Framework to Understand Sustainable Technological Change: Empirical Evidence for Earlier Claims. *Technological Forecasting and Social Change* 76 (4): 584–594.

Henning, Kroll. 2011. Exploring the Validity of Patent Applications as an Indicator of Chinese Competitiveness and Market Structure. *World Patent Information* 33 (1): 23–33.

Hirsh, Richard F. 1999. PURPA: The Spur to Competition and Utility Restructuring. *Electricity Journal* 12 (7): 60–72.

Holdren, John P., and Kirk Smith. 2000. Energy, the Environment, and Health. In *World Energy Assessment: Energy and the Challenge of Sustainability*, ed. José Goldemberg. New York: UN Development Program.

Hook, Leslie. 2012. AMSC Dispute Goes to China's Supreme Court. *Financial Times*, April 9.

Howarth, Robert W., Renee Santoro, and Anthony Ingraffea. 2011. Methane and the Greenhouse-Gas Footprint of Natural Gas from Shale Formations. *Climatic Change* 106 (4): 679–690.

Hu, Albert Guangzhou. 2010. Propensity to Patent, Competition, and China's Foreign Patenting Surge. *Research Policy* 39 (7): 985–993.

Hu, Albert Guangzhou, and Gary H. Jefferson. 2009. A Great Wall of Patents: What Is Behind China's Recent Patent Explosion? *Journal of Development Economics* 90 (1): 57–68.

IEA (International Energy Agency). 2001. Technology without Borders: Case Studies of Successful Technology Transfer. http://www.climatetech.net/pdf/Ctifull .pdf.

IEA (International Energy Agency). 2009. *World Energy Outlook*. Paris: International Energy Agency, Organization for Economic Cooperation and Development.

IEA (International Energy Agency). 2013. Energy Subsidies. http://www .worldenergyoutlook.org/resources/energysubsidies (accessed October 21, 2013).

IEA/IRENA. 2012. Global Renewable Energy Policies and Measures Database. OECD/IEA and International Renewable Energy Agency. http://www.iea.org/ textbase/pm/index.html (accessed August 22, 2012).

Inoue, Yuko. 2010. China Lifts Rare Earth Export to Japan. Reuters, September 29.

International Institute for Sustainable Development. 2006. Climate Technology Initiative Industry Joint Seminar on Successful Cases of Tech Transfer in Asian Countries. http://www.iisd.ca/download/pdf/sd/ymbvol92num5e.pdf (accessed July 9, 2013).

IPC Committee of Experts. 2012. IPC Green Inventory. Geneva: World Intellectual Property Organization. http://www.wipo.int/classifications/ipc/en/est (accessed March 5, 2012).

IPCC (Intergovernmental Panel on Climate Change). 2000. *Methodological and Technological Issues in Technology Transfer*, ed. Bert Metz, Ogunlade Davidson, Jan-Willem Martens, Sascha van Rooijen, and Laura van wie Mcgrory. Cambridge: Cambridge University Press.

IPCC. 2005. *Special Report on Carbon Dioxide Capture and Storage*. Cambridge, UK: Cambridge University Press.

IPCC. 2012. Technical Summary. In *Renewable Energy Sources and Climate Change Mitigation: Special Report of the Intergovernmental Panel on Climate Change*, ed. Ottmar Edenhofer, Ramon P. Madruga, and Youba Sokona. Cambridge: Cambridge University Press.

IRENA. 2012a. Solar Photovoltaics; Renewable Energy Technologies: Cost Analysis Series. *International Renewable Energy Agency* 1 (4–5).

IRENA. 2012b. Wind Power; Renewable Energy Technologies: Cost Analysis Series. *International Renewable Energy Agency* 1 (5–5).

Ji, Shuguang, Christopher Cherry, Matthew Bechle, Ye Wu, and Julien Marshall. 2012. EV's in China: Emissions and Health Impacts. *Environmental Science and Technology* 46:2018–2024.

Jia, Heping. 2011. Chinese Science Ministry Reveals Budget. http://blogs.nature.com/news/2011/04/chinese_science_ministry_revea.html (accessed May 11, 2012).

Jie, Ma and Yuki Hagiwara. 2013. In Ghosn We Trust Tested as Nissan Electric Push Falters. Bloomberg, March 21.

Johnson, Miles. 2013. Green Energy Hit by Spanish Reform. *Financial Times*, February 7.

Karnoe, Peter, and Adam Buchhorn. 2008. Denmark: Path-Creation Dynamics and Winds of Change. In *Promoting Sustainable Electricity in Europe: Challenging the Path Dependence of Dominant Energy Systems*, ed. William M. Lafferty and Audun Ruud, 73–101. Cheltenham, UK: Edward Elgar.

Kemp, René. 1997. *Environment Policy and Technical Change*. Cheltenham, UK: Edward Elgar.

Kemp, René, Johan Schot, and Remco Hoogma. 1998. Regime Shifts to Sustainability through Processes of Niche Formation: The Approach of Strategic Niche Management. *Technology Analysis and Strategic Management* 10 (2): 175–196.

Kempener, Rudd, Laura Diaz Anadon, Jose Condor, and Joel Kenrick. 2010. Governmental Energy Innovation Investments, Policies, and Institutions in the Major Emerging Economies: Brazil, Russia, India, Mexico, China, and South Africa. Cambridge, MA: Belfer Center for Science and International Affairs, Harvard University.

Kobos, Peter H., Jon D. Erickson, and Thomas E. Drennen. 2006. Technological Learning and Renewable Energy Costs: Implications for US Renewable Energy Policy. *Energy Policy* 34 (13): 1645–1648.

Kraemer, Susan. 2012. Japan Creates Potential $9.6 Billion Boom with FITS. http://cleantechnica.com (accessed June 21, 2012).

Kristinsson, Kári, and Rekha Rao. 2008. Interactive Learning or Technology Transfer as a Way to Catch Up? Analyzing the Wind Energy Industry in Denmark and India. *Industry and Innovation* 15 (3): 297–320.

Krueger, Anne O. 1974. The Political Economy of the Rent-Seeking Society. *American Economic Review* 64 (3): 291–303.

Krugman, Paul. 1997. *Development, Geography, and Economic Theory*. Paperback ed. Cambridge, MA: MIT Press.

Lane, Eric L. 2013. *Clean Tech Intellectual Property: Eco-marks, Green Patents, and Green Innovation*. Danvers, MA: LexisNexis.

Lanjouw, Jean Olson, and Ashoka Mody. 1996. Innovation and the International Diffusion of Environmentally Responsive Technology. *Research Policy* 25 (4): 549–571.

Lappin, Joan. 2011. American Superconductor Destroyed for a Tiny Bribe. *Forbes* (September): 21.

Lau, Lee Chung, Kok Tat Tan, Keat Teong Lee, and Abdul Rahman Mohamed. 2009. A Comparative Study on the Energy Policies in Japan and Malaysia in Fulfilling Their Obligations towards the Kyoto Protocol. *Energy Policy* 37 (11): 4771–4778.

Lee, Bernice, Ilian Illiev, and Felix Preston. 2009. *Who Owns Our Low Carbon Future? Intellectual Property and Energy Technologies*. London: Chatham House.

Lema, Adrian, and Kristian Ruby. 2007. Between Fragmented Authoritarianism and Policy Coordination: Creating a Chinese Market for Wind Energy. *Energy Policy* 35 (July): 3879–3890.

Lema, Rasmus, and Adrian Lema. 2010. Whither Technology Transfer? The Rise of China and India in Green Technology Sectors. Unpublished manuscript.

Lester, Richard K., and Michael J. Piore. 2004. *Innovation: The Missing Dimension*. Cambridge, MA: Harvard University Press.

Levi, Michael A., Elizabeth C. Economy, Shannon K. O'Neil, and Adam Segel. 2010. Energy Innovation: Driving Technology Competition and Cooperation among US, China, India, Brazil. http://www.cfr.org/india/energy-innovation/p23321 (accessed July 9, 2013).

Lewis, Joanna I. 2007. Technology Acquisition and Innovation in the Developing World: Wind Turbine Development in China and India. *Studies in Comparative International Development* 42 (3–4): 208–232.

Lewis, Joanna I., 2011. Building a National Wind Turbine Industry: Experiences from China, India, and South Korea. *International Journal of Technology Transfer and Commercialization* 5 (3–4): 281–305.

Lewis, Joanna I. 2013. *Green Innovation in China: China's Wind Power Industry and the Global Transition to a Low Carbon Economy*. New York: Columbia University Press.

Lewis, Joanna I., and Ryan H. Wiser. 2007. Fostering a Renewable Energy Technology Industry: An International Comparison of Wind Industry Policy Support Mechanisms. *Energy Policy* 35 (3): 1844–1857.

Lewis, Nathan S. 2007. Toward Cost-Effective Solar Energy Use. *Science* 315 (5813): 798–801.

Li, Junfeng, Pengfei Shi, and Gao Hu. 2010. *2010 China Wind Power Outlook*. Brussels: Global Wind Energy Council.

Li, Xibao. 2012. Behind the Recent Surge of Chinese Patenting: An Institutional View. *Research Policy* 41 (1): 236–249.

Li, Zheng. 2012. Perspectives on Energy in China: An Update through 2011. Paper presented at the Carbon Mitigation Initiative annual meeting, Princeton, NJ.

Liebert, Tilmann. 2011. Competitiveness of Renewable Energies in Climate Change Policy: Explaining Post-Kyoto Emission Reduction Commitments. Master's thesis, Fletcher School, Tufts University.

Lin, Boqiang, and Xuehui Li. 2011. The Effect of Carbon Tax on Per Capita CO_2 Emissions. *Energy Policy* 39 (9): 5137–5146.

Lin, Justin Yifu. 2012. *Demystifying the Chinese Economy*. Cambridge: Cambridge University Press.

Lipp, Judith. 2007. Lessons for Effective Renewable Electricity Policy from Denmark, Germany, and the United Kingdom. *Energy Policy* 35:5481–5495.

Liu, Hengwei, and Kelly Sims Gallagher. 2010. Catalyzing Strategic Transformation to a Low-Carbon Economy: A CCS Roadmap for China. *Energy Policy* 38 (1): 59–74.

Liu, Li-qun, Zhi-xin Wang, Hua-qiang Zhang, and Ying-cheng Xue. 2010. Solar Energy Development in China—A review. *Renewable and Sustainable Energy Reviews* 14 (1): 301–11.

Liu, Yingqi, and Ari Kokko. 2010. Wind Power in China: Policy and Development Challenges. *Energy Policy* 38 (October): 5520–5529.

Lovins, Amory, and L. Hunter Lovins. 1982. *Brittle Power: Energy Strategy for National Security*. Baltimore: Brick House Publishing.

Lu, Zifeng, Qiang Zhang, and David G. Streets. 2011. Sulfur Dioxide and Primary Carbonaceous Aerosol Emissions in China and India, 1996–2010. *Atmospheric Chemistry and Physics* 11:9839–9864.

Lundvall, Bengt-Åke. 1988. Innovation as an Interactive Process: From User-Producer Interaction to the National System of Innovation. In *Technical Change and Economic Theory*, ed. Giovanni Dosi, Christopher Freeman, Richard R. Nelson, Gerald Silverberg, and Luc L. Soete, 349–369. London: Pinter Publishers.

Lundvall, Bengt-Åke. 2007. National Innovation Systems-Analytical Concept and Development Tool. *Industry and Innovation* 14 (1): 95–119.

Lundvall, Bengt-Åke. 2009. Innovation as an Interactive Process: User-Producer Interaction to the National System of Innovation. *African Journal of Science, Technology, Innovation, and Development* 1 (2–3): 10–34.

Mabuza, Lindiwe O. K., Alan C. Brent, and Maxwell Mapako. 2007. The Transfer of Energy Technologies in a Developing Country Context—Toward Improved Practice from Past Successes and Failures. Proceedings of World Academy of Science, Engineering, and Technology 22 (July).

Marigo, Nicoletta, T. J. Foxon, and Peter J. Pearson. 2008. Comparing Innovation Systems for Solar Photovoltaics in the United Kingdom and China. Unpublished manuscript, DSpaceUnipr.

Martinot, Eric, Jonathan E. Sinton, and Brent M. Haddad. 1997. International Technology Transfer for Climate Change Mitigation, and the Cases of Russia and China. *Annual Review of Energy and the Environment* 22:357–401.

Maskus, Keith E. 2000. *Intellectual Property Rights in the Global Economy.* Washington, DC: Institute for International Economics.

Maskus, Keith E., and Ruth L. Okediji. 2010. Intellectual Property Rights and International Technology Transfer to Address Climate Change: Risks, Opportunities, and Policy Options. Geneva: International Centre for Trade and Sustainable Development.

Matus, Kira, Kyung-Min Nam, Noelle E. Selin, Lok N. Lamsal, John M. Reilly, and Sergey Paltsev. 2012. Health Damages from Air Pollution in China. *Global Environmental Change* 22 (1): 55–66.

McCrone, Angus. 2012. *Global Trends in Renewable Energy Investment 2012.* Frankfurt: Frankfurt School of Finance and Management, UN Environment Program, Bloomberg New Energy Finance.

McDonald, Joe. 2012. China's Dream of Electric Car Leadership Elusive. Associated Press, April 24.

Mertha, Andrew. 2005. *The Politics of Piracy: Intellectual Property in Contemporary China.* Ithaca, NY: Cornell University Press.

Metcalf, Gilbert, and David Weisbach. 2009. The Design of a Carbon Tax. *Harvard Environmental Law Review* 33 (2): 499–556.

Metzger, Eliot, comp. 2008. Bottom Line on Cap-and-Trade. World Resources Institute, July. http://www.wri.org/publication/bottom-line-cap-and-trade (accessed July 8, 2012).

Ministry of Environmental Protection (China). 2012. http://english.mep.gov.cn (accessed July 2, 2012).

MIT. 2012. GreenGen Fact Sheet. MIT Energy Initiative. Sequestration.mit.edu/tools/projects/greengen.html.

Mizuno, Emi. 2007. Cross-Border Transfer of Climate Change Mitigation Technologies: The Case of Wind Energy from Denmark and Germany to India. PhD diss., Massachusetts Institute of Technology.

Mohammad, Ali. 2001. PhD diss., Massey University. http://www.mro.massey.ac.nz/handle/10179/2097 (accessed July 8, 2013).

Monge, Karol Acon. 2000. Barriers to Technology Transfer for Climate Change Mitigation: The Case of the Indian Coal-Fired Thermal Power Industry. http://www.devalt.org/newsletter/jun00/of_6.htm (accessed July 9, 2013).

Monkelbaan, Joaquim. 2011. Trade Preferences for Environmentally Friendly Goods and Services. Geneva: Global Platform on Climate Change, Trade, and Sustainable Energy, International Centre for Trade and Sustainable Development.

Moomaw, William R. 2012. Personal communication, Medford, MA.

Mowrey, David, and Joanne E. Oxley. 1997. Inward Technology Transfer and Competitiveness: The Role of National Innovation Systems. In *Technology, Globalization, and Economic Performance*, ed. Daniele Archibugi and Jonathan Michie, 138–171. Cambridge: Cambridge University Press.

Murkowski, Lisa. 2011. Hearing Statement. Senate Energy and Natural Resources Committee, Washington, DC, March 17.

Murphy, James T. 2001. Making the Energy Transition in Rural East Africa: Is Leapfrogging an Alternative? *Technological Forecasting and Social Change* 68 (2): 173–193.

Nam, Kyung-Min. 2011. Learning through the International Joint Venture: Lessons from the Experience of China's Automotive Sector. *Industrial and Corporate Change* 20 (3): 855–907.

NASA. 2005. China's Wall Less Great in View from Space, in NASA. http://www.nasa.gov/vision/space/workinginspace/great_wall.html (accessed April 12, 2012).

Naughton, Barry. 2007. *The Chinese Economy: Transitions and Growth*. Cambridge, MA: MIT Press.

National Bureau of Statistics. 2011. *China Statistical Database*. Beijing: National Bureau of Statistics of China.

Neij, Lena. 1999. Cost Dynamics of Wind Power. *Energy* 24 (5): 375–389.

Neij, Lena. 2008. Cost Development of Future Technologies for Power Generation: A Study Based on Experience Curves and Complementary Bottom-Up Assessments. *Energy Policy* 36 (6): 2200–2211.

Nelson, Richard R., ed. 1993. *National Innovation Systems: A Comparative Analysis*. New York: Oxford University Press.

Nelson, Richard R., and Sidney G. Winter. 1982. *An Evolutionary Theory of Economic Change*. Cambridge, MA: Harvard University Press.

Nemet, Gregory F. 2006. Beyond the Learning Curve: Factors Influencing Cost Reductions in Photovoltaics. *Energy Policy* 34 (17): 3218–3232.

Nemet, Gregory F. 2009. Demand-Pull, Technology-Push, and Government-Led Incentives for Non-Incremental Technical Change. *Research Policy* 38 (5): 700–709.

Neubacher, Alexander. 2012. A Capital Error? Germany Created Its Own Threat with Chinese Solar Aid. *Spiegel*, February 27.

Nguyen, Nhan T., Minh Ha-Duong, Thanh C. Tran, Ram M. Shrestha, and Franck Nadaud. 2010. Barriers to the Adoption of Renewable and Energy-Efficient Technologies in the Vietnamese Power Sector. http://minh.haduong.com/files/Nguyen.ea-20100120-BarriersVietnam.pdf (accessed July 9, 2013).

Nordhaus, William D. 2004. Schumpeterian Profits in the American Economy: Theory and Management. No. 10433. Cambridge, MA: National Bureau of Economic Research.

Nordqvist, Joakim. 2006. Evaluation of Japan's Top Runner Programme. EIE-2003-114. Europe: Active Implementation of the European Directive on Energy Efficiency.

NREL. 2012. Sunshot Vision Study. National Renewable Energy Laboratory, US Department of Energy, Washington, DC. http://www1.eere.energy.gov/pdfs/47927.pdf.

Obama, Barack. 2011. Energy, Climate Change, and Our Environment. Washington, DC: White House. http://www.whitehouse.gov/energy (accessed July 27, 2011).

Ockwell, David G. 2012. Adapting Energy Policy for Climate Change: Challenges and Opportunities for LDCs, SVEs, and SIDs. In *Low-Carbon Technology Transfer: From Rhetoric to Reality*, ed. David G. Ockwell and Alexandra Mallett. Abingdon, UK: Earthscan.

Ockwell, David G., Ruediger Haum, Alexandra Mallett, and Jim Watson. 2010. Intellectual Property Rights and Low Carbon Technology Transfer: Conflicting Discourses of Diffusion and Development. *Global Environmental Change* 20 (4): 729–738.

Ockwell, David G., and Alexandra Mallett, eds. 2012. *Low Carbon Technology Transfer: From Rhetoric to Reality*. Abingdon, UK: Routledge.

Ockwell, David G., Jim Watson, Gordon MacKerron, Prosanto Pal, and Farhana Yamin. 2008. Key Policy Considerations for Facilitating Low Carbon Technology Transfer to Developing Countries. *Energy Policy* 36:4104–4115.

Odagiri, Hiroyuki, Akira Goto, Atsushi Sunami, and Richard R. Nelson. 2010. Introduction to *Intellectual Property Rights, Development, and Catch Up: An International Comparative Study*, ed. Hiroyuki Odagiri, Akira Goto, Atsushi Sunami, and Richard R. Nelson. Oxford: Oxford University Press.

Oliver, Hongyan H., Kelly Sims Gallagher, Donglian Tian, and Jinhua Zhang. 2009. China's Fuel Economy Standards for Passenger Vehicles: Rationale, Policy Process, and Impacts. *Energy Policy* 37 (11): 4720–4729.

Osborne, Mark. 2011. Top Executives from 14 Leading Chinese PV Manufacturers Gather to Fight SolarWorld. PV Tech. November 30.

Pachauri, R. K., and Preety Bhandari, eds. 1994. *Climate Change in Asia and Brazil: The Role of Technology Transfer*. New Delhi: Tata Energy Research Institute.

Pal, Prosanto, and Girish Sethi. 2012. Technology Transfer of Energy-Efficient Technologies among Small and Medium Enterprises in India. *Low Carbon Technology Transfer: From Rhetoric to Reality*, ed. David G. Ockwell and Alexandra Mallett. Abingdon, UK: Earthscan.

Park, Walter G., and Douglas Lippoldt. 2005. International Licensing and the Strengthening of Intellectual Property Rights in Developing Countries During the 1990s. *OECD Economic Studies* 40.

Patel, Surendra J., Pedro Roffe, and Abdulqawi Yusuf, eds. 2001. *International Technology Transfer: The Origins and Aftermath of the United Nations Negotiations on a Draft Code of Conduct.* The Hague: Kluwer Law International.

Pauschert, Dirk. 2009. Study of Equipment Prices in the Power Sector. ESMAP Technical Paper 122/09. Washington, DC: International Bank for Reconstruction and Development.

Pavitt, Keith. 1982. R&D, Patenting, and Innovative Activities: A Statistical Exploration. *Research Policy* 11 (1): 33–51.

People's Bank of China. 2010. Survey and Statistics from the People's Bank of China. Beijing: People's Bank of China. http://www.pbc.gov.cn/publish/english/958/index.html (accessed May 10, 2012).

People's Daily. 2009. Central Government to Subsidize Foreign Patent Applications. *People's Daily*, October 13.

Pew Charitable Trusts. 2012. *Who's Winning the Clean Energy Race?* Washington, DC: Pew Charitable Trusts.

Pew Charitable Trusts. 2013. Who's Winning the Clean Energy Race, 2012? http://www.pewenvironment.org/uploadedFiles/PEG/Publications/Report/-clenG20-Report-2012-Digital.pdf (accessed July 3, 2013).

Pfeifer, Sylvia. 2012. Finds That Form a Bedrock of Hope. *Financial Times*, April 23.

Popp, David C. 2001. Data Appendix for "The Effect of New Technology on Energy Consumption." http://faculty.maxwell.syr.edu/dcpopp/papers/dataapnd.pdf (accessed March 5, 2012).

Porter, Eduardo. 2012. China's Vanishing Trade Imbalance. *New York Times*, May 2, B1.

Porter, Michael E., and Claas van der Linde. 1995. Toward a New Conception of the Environment-Competitiveness Relationship. *Journal of Economic Perspectives* 9 (4): 97–118.

Price, Lynn, Mark D. Levine, Nan Zhou, David Fridley, Nathaniel Aden, Hongyou Lu, Michael McNeil, Nina Zheng, Yining Qin, and Ping Yowargana. 2011. Assessment of China's Energy-Saving and Emission-Reduction Accomplishments and Opportunities during the 11th Five-Year Plan. *Energy Policy* 39 (4): 2165–2178.

PriceWaterhouseCoopers China Pharmaceutical Team. 2009. Investing in China's Pharmaceutical Industry. Beijing: PriceWaterHouseCoopers.

Ramsey, Mike. 2012. China's New Target: Batteries. *Wall Street Journal*, August 12.

Ratliff, Phil, Paul Garbett, and Willibald Fischer. 2007. *The New Siemens Gas Turbine SGT5-8000H for More Customer Benefit.* Essen, Germany: VGB PowerTech.

Reddy, N. Mohan, and Liming Zhao. 1990. International Technology Transfer: A Review. *Research Policy* 19:285–307.

RE4. 2011. Renewable Energy and Energy Efficiency Export Initiative on Track to Increase RE&EE Exports. Washington, DC: National Export Initiative, US Department of Commerce.

RE4 Advisory Committee. 2012. *Meeting Minutes for June 14, 2012*. Washington, DC: Renewable Energy and Energy Efficiency Advisory Committee, US Department of Commerce.

Regional GHG Initiative. 2012. Old Auctions. http://www.rggi.org/market/co2 _auctions/information/old_auction_notices (accessed 19 September 2012).

Ren21. 2012. Renewables 2012: Global Status Report. Paris: Renewable Energy Policy Network.

Rickerson, Wilson, Janet Sawin, and Robert Grace. 2007. If the Shoe FITs: Using Feed In Tariffs to Meet US Renewable Energy Targets. *Electricity Journal* 20 (4): 73–86.

Riley, Michael, and Ashley Vance. 2012. China Corporate Espionage Boom Knocks Wind Out of US Companies. *Bloomberg Businessweek*, March 15.

Rock, Michael T., and David P. Angel. 2005. *Industrial Transformation in the Developing World*. Oxford: Oxford University Press.

Rogers, Everett M. 1995. *Diffusion of Innovations*. New York: Free Press.

Ru, Peng, Qiang Zhi, Fang Zhang, Xiaotian Zhong, Jianqiang Li, and Jung Su. 2012. Behind the Development of Technology: The Transition of Innovation Modes in China's Wind Turbine Manufacturing Industry. *Energy Policy* 43:58–69.

Sauter, Raphael, and Jim Watson. 2008. *Technology Leapfrogging: A Review of the Evidence*. Sussex: University of Sussex.

Sawin, Janet. 2001. The Role of Government in the Development and Diffusion of Renewable Energy Technologies: Wind Power in the United States, California, Denmark, and Germany, 1970–2000. PhD diss., Fletcher School of Law and Diplomacy, Tufts University.

Sawin, Janet. 2012. *Renewables 2011: Global Status Report*. Paris: Renewable Energy Policy Network.

SBS Staff. 2012. Factbox: Carbon Taxes around the World. World News Australia, June 27. http://www.sbs.com.au/news/article/1492651/factbox-carbon-taxes -around-the-world (accessed July 8, 2012).

Scherer, Frederic M., Sigmund E. Herzstein, Alex Dreyfoos, William Whitney, Otto Bachmann, Paul Pesek, Charles Scott, Thomas Kelly, and James Galvin. 1959. Patents and the Corporation: A Report on Industrial Technology under Changing Public Policy. Boston: Patents and the Corporation.

Schmookler, Jacob. 1966. *Invention and Economic Growth*. Cambridge, MA: Harvard University Press.

Schoots, Koen, Francesco Ferioli, Gert Jan Kramer, and Bob C. C. van der Zwaan. 2008. Learning Curves for Hydrogen Production Technology: An Assessment of Observed Cost Reductions. *International Journal of Hydrogen Energy* 33 (11): 2630–2645.

Schumacher, Ernst F. 1973. *Small Is Beautiful: Economics as If People Mattered.* New York: Harper and Row.

Schumpeter, Joseph A. 2004. *The Theory of Economic Development.* Livingston, NJ: Transaction Publishers. [Originally published in 1934 by President and Fellows of Harvard University, Cambridge, MA.]

Searchinger, Timothy, Ralph Heimlich, R. A. Houghton, Fengxia Dong, Amani Elobeid, Jacinto Fabiosa, Simia Tokgoz, Dermot Hayes, and Tun-Hsiang Yu. 2008. Use of U.S. Croplands for Biofuels Increases Greenhouse Gases through Emissions from Land-Use Change. *Science* 319 (5867): 1238–1240.

Seebregts, Ad J. 2010. Gas-Fired Power. Paris: International Energy Agency.

SIPO. 2010. Guidelines for Examination. State Intellectual Property Office of People's Republic of China. http://www.sipo.gov.cn/zlsqzn/sczn2010eng.pdf.

SIPO. 2012. Statistics. State Intellectual Property Office of People's Republic of China. http://www.english.sipo.gov/statistics/2012/1.

Smil, Vaclav. 2010. *Prime Movers of Globalization: The History and Impact of Diesel Engines and Gas Turbines.* Cambridge, MA: MIT Press.

Soete, Luc L. 1985. International Diffusion of Technology, Industrial Development, and Technological Leapfrogging. *World Development* 13 (3): 409–422.

SolarBuzz. 2012. Module Pricing: Retail Price Summary, March Update. http://www.solarbuzz.com/facts-and-figures/retail-price-environment/module-prices (accessed September 9, 2012).

State Council Information Office. 2007. *Environmental Protection in China: 1996–2005.* Beijing: State Council Information Office.

Steinfeld, Edward S. 2010. *Playing Our Game: Why China's Rise Doesn't Threaten the West.* Oxford: Oxford University Press.

Stern, Nicholas. 2006. *Stern Review: The Economics of Climate Change.* Cambridge: Cambridge University Press.

Sumner, Jenny, Lori Bird, and Hillary Smith. 2009. Carbon Taxes: A Review of Experience and Policy Design Considerations. Technical Report NREL/TP-6A2-47312. December. Golden, CO: National Renewable Energy Laboratory. http://www.nrel.gov/docs/fy10osti/47312.pdf (accessed July 19, 2013).

Supreme People's Court. 2010. *Intellectual Property Protection by Chinese Courts in 2009.* Beijing: Supreme People's Court, People's Republic of China.

Suttmeier, Richard P., and Xiangjui Yao. 2011. *China's IP Transition: Rethinking Intellectual Property Rights in a Rising China.* Seattle: National Bureau of Asian Research.

Tan, Chunrong. 2012. Analysis of China's Oil and Gas Imports and Exports in 2011. *International Petroleum Economics Monthly* 3:56–66.

Taylor, Christopher T., and Z. Aubrey Silberston. 1973. *The Economic Impact of the Patent System.* Cambridge: Cambridge University Press.

Taylor, Margaret. 2008. Beyond Technology-Push and Demand-Pull: Lessons from California's Solar Policy. *Energy Economics* 30 (6): 2829–2854.

Teece, David J. 1977. Technology Transfer by Multinational Firms: The Resource Cost of Transferring Technological Know-How. *Economic Journal* 87 (346): 242–261.

TERI. 2000. Lessons from Case Studies of Tech Transfer of Climate Technologies in the Asian Region. The Energy Research Institute. New Dehli. http://www .climatetech.net/template.cfm?FrontID=3942 (accessed July 9, 2013).

Trachtman, Joel P. 2009. *The International Law of Economic Migration: Toward the Fourth Freedom*. Kalamazoo, MI: W. E. Upjohn Institute for Employment Research.

Tu, Jianjun. 2011. Industrial Organization of the Chinese Coal Industry. Stanford, CA: Stanford Program on Energy and Sustainable Development.

Ueno, Takahiro. 2009. Technology Transfer to China to Address Climate Change Mitigation. Resources for the Future Issue Brief. http://www.rff.org/RFF/ Documents/RFF-IB-09-09.pdf (accessed July 9, 2013).

UN Educational, Scientific, and Cultural Organization. 2011. *Global Education Digest*. Montreal: UNESCO Institute for Statistics.

UNCTAD-ICTSD. 2005. *Resource Book on TRIPS and Development*. Cambridge: Cambridge University Press.

UNEP, EPO, and ICTSD. 2010. Patents and Clean Energy: Bridging the Gap between Evidence and Policy. Geneva: UN Environment Program, European Patent Office, and International Centre for Trade and Sustainable Development.

Unruh, Gregory C. 2000. Understanding Carbon Lock-In. *Energy Policy* 28 (12): 817–830.

US Census. 2012. *U.S. Imports of Crude Oil*. Washington, DC: US Census Bureau, Foreign Trade Division.

US EPA (US Environmental Protection Agency). 2010. Information on Compact Fluorescent Lightbulbs and Mercury (FAQs). Washington, DC: US Environmental Protection Agency. http://www.energystar.gov/ia/partners/promotions/ change_light/downloads/Fact_Sheet_Mercury.pdf (accessed July 26, 2011).

US EPA. 2012. History of the Clean Air Act. Last modified February 17, 2012. Accessed September 13, 2012. http://epa.gov/oar/caa/caa_history.html.

US Trade Representative. 2011. Press Release: U.S. and China Conclude 22nd Session of the Joint Commission on Commerce and Trade. Office of the US Trade Representative, Washington, DC, November.

Vernon, Raymond, and William H. Davidson. 1979. Foreign Production of Technology-Intensive Products by US-Based Multinational Enterprises. Working Paper, Graduate School of Business Administration, Harvard University, Cambridge, MA.

Vollebergh, Herman R.J. 2008. Lessons from the Polder: Energy Tax Design in the Netherlands from a Climate Change Perspective. *Ecological Economics* 64 (3): 660–672. http://dx.doi.org.ezproxy.library.tufts.edu/10.1016/j.ecolecon .2007.04.011 (accessed July 9, 2013).

Walz, Rainer. 2010. Competences for Green Development and Leapfrogging in Newly Industrializing Countries. *International Economics and Economic Policy* 7 (2–3): 245–265.

Wang, Bing. 2007. An Imbalanced Development of Coal and Electricity Industries in China. *Energy Policy* 35:4959–4968.

Wang, Bo. 2010. Can CDM Bring Technology Transfer to China? An Empirical Study of Technology Transfer in China's CDM. *Energy Policy* 38 (5): 2572–2585.

Wang, Hua, and Chris Kimble. 2011. Leapfrogging to Electric Vehicles: Patterns and Scenarios for China's Automobile Industry. *International Journal of Automotive Technology and Management* 11 (4): 312–325.

Wang, Lifang. 2013. Personal communication, Chinese Academy of Sciences.

Warhurst, Alyson. 1991. Technology Transfer and the Development of China's Offshore Oil Industry. *World Development* 19 (8): 1055–1073.

Watal, Jayashree. 2007. The Issue of Technology Transfer in the Context of the Montreal Protocol: Case Study of India. In *Achieving Objectives of Multilateral Environmental Agreements: Lessons from Empirical Studies*, ed. Veena Jha and Ulrich Hoffmann. UNCTAD/ITCD/TED/6.

Watanabe, Chihiro Kouji Wakabayashi, and Toshinori Miyazawa. 2000. Industrial Dynamism and the Creation of a "Virtuous Cycle" between R&D, Market Growth, and Price Reduction: The Case of Photovoltaic Power Generation (PV) Development in Japan. *Technovation* 20 (6): 299–312.

Watson, Jim. 2004. Selection Environments, Flexibility, and the Success of the Gas Turbine. *Research Policy* 33 (8): 1065–1080.

Watson, Jim, Rob Byrne, David G. Ockwell, Michele Stua, and Alexandra Mallett. 2010. Low Carbon Technology Transfer: Lessons from India and China. Sussex, UK: Tyndall Centre for Climate Change Research, University of Sussex. http://www.tyndall.ac.uk/sites/default/files/Briefing_Note_45.pdf (accessed January 1, 2012).

Watson, Jim, Robert Byrne, Michele Stua, David G. Ockwell, and Xiliang Zhang. Da Zhang, Tianhou Zhang, Xiaofeng Zhang, Xunmin Ou, and Alexandra Mallett. 2011. UK-China Collaborative Study on Low Carbon Technology Transfer: Final Report to the Department of Energy and Climate Change. Sussex Energy Group. http://www.sussex.ac.uk/sussexenergygroup/research/growthinnovationdevelopingcountries/Ukindiacollaberationjim (accessed July 9, 2013).

Watson, Jim, and Geoffrey Oldham. 2000. International Perspectives on Clean Coal Technology Transfer to China: Final Report to the Working Group on Trade and Environment. Sussex, UK, August.

Watts, Jennifer, and Kyle Bagin. 2010. *Critical Technology Assessment: Impact of US Export Controls on Green Technology Items*. Washington, DC: Bureau of Industry and Technology.

Webb, Tim. 2009. Shell Dumps Wind, Solar, and Hydro Power in Favour of Biofuels. *Guardian*, March 17.

Weir, Mette, Katja Birr-Pedersen, Henrik Klinge Jacobsen, and Jacob Klok. 2005. Are CO_2 Taxes Regressive? Evidence from the Danish Experience. *Ecological Economics* 52: 239–251.

Weismann, Gerrit. 2012. Berlin to Reduce Solar Power Payouts. *Financial Times*, February 23.

Weiss, Richard. 2012. Siemens Pumps $1.3 Billion into Gas Turbines to Fend Off GE. Bloomberg, January 12.

Weitzman, Martin L. 2009. On Modeling and Interpreting the Economics of Catastrophic Climate Change. *Review of Economics and Statistics* 91 (1): 1–19.

Wen Jiabao. 2009. Speech by H. E. Wen Jiabao, premier, People's Republic of China, at the fifth China-EU Business Summit, Nanjing, November 30.

Wen Jiabao. 2011. National Teleconference on the Work of Saving Energy and Reducing Emissions. Accessed through World News Connection, Xinhua Domestic Service, October 2.

Wesoff, Eric. 2012. Consolidation Chronicles: Q-Cells Has a Suitor, Global Solar Unit for Sale, REC Shutdown: Another Week, Another Seismic Shift in the Solar Industry. Greentechmedia.com, April 25.

Wilkins, Gill. 2002. *Technology Transfer for Renewable Energy: Overcoming Barriers for Developing Countries*. London: Royal Institute of International Affairs.

Wilson, Charlie. 2012. Upscaling, Formative Phases, and Learning in the Historical Diffusion of Energy Technologies. *Energy Policy* 50: 81–94.

Wong, Christine. 2009. Rebuilding Government in the 21st Century: Can China Incrementally Reform the Public Sector? *China Quarterly* 200 (December): 929–952.

World Bank. 2012. Databank. http://databank.worldbank.org/ (accessed June 27, 2013).

World Bank and Development Research Center of the State Council. 2012. *China 2030: Building a Modern, Harmonious, and Creative High-Income Society*. Washington, DC: World Bank.

World Bank, State Environmental Protection Administration of China. 2007. *Cost of Pollution in China: Economic Estimates of Physical Damages*. Washington, DC: World Bank.

Worldwatch Institute. 2008. *Green Jobs: Toward Decent Work in a Sustainable Low Carbon World*. Nairobi: UN Environment Program.

Xuan, Xiaowei, and Kelly Sims Gallagher. 2013. Prospects for Reducing Carbon Intensity in China. Center for International Environment and Resource Policy Discussion Paper. Medford, MA: Fletcher School, Tufts University.

Yang, Hong, He Wang, Huacong Yu, Jianping Xi, Rongqiang Cui, and Guangde Chen. 2003. Status of Photovoltaic Industry in China. *Energy Policy* 31 (8): 703–707.

Zhao, Lifeng, and Kelly Sims Gallagher. 2007. Research, Development, Demonstration, and Early Deployment Policies for Advanced-Coal Technology in China. *Energy Policy* 35 (12): 6467–6477.

Zhao, Lifeng, Yunhan Xiao, Kelly Sims Gallagher, Bo Wang, and Xiang Xu. 2008. Technical, Environmental, and Economic Assessment of Deploying Advanced Coal Power Technologies in the Chinese Context. *Energy Policy* 36 (7): 2709–2718.

Zhao, Ruirui, Guang Shi, Hongyu Chen, Anfu Ren, and David Finlow. 2011. Present Status and Prospects of Photovoltaic Market in China. *Energy Policy* 39 (4): 2204–2207.

Zhou, Eve Y., and Bob Stembridge. 2008. *Patented in China: The Present and Future State of Innovation in China*. Philadelphia: Thompson Reuters.

Index

Urban and Industrial Environments

Series editor: Robert Gottlieb, Henry R. Luce Professor of Urban and Environmental Policy, Occidental College

Anastasia Loukaitou-Sideris and Renia Ehrenfeucht, *Sidewalks: Conflict and Negotiation over Public Space*

David J. Hess, *Localist Movements in a Global Economy: Sustainability, Justice, and Urban Development in the United States*

Julian Agyeman and Yelena Ogneva-Himmelberger, eds., *Environmental Justice and Sustainability in the Former Soviet Union*

Jason Corburn, *Toward the Healthy City: People, Places, and the Politics of Urban Planning*

JoAnn Carmin and Julian Agyeman, eds., *Environmental Inequalities beyond Borders: Local Perspectives on Global Injustices*

Louise Mozingo, *Pastoral Capitalism: A History of Suburban Corporate Landscapes*

Gwen Ottinger and Benjamin Cohen, eds., *Technoscience and Environmental Justice: Expert Cultures in a Grassroots Movement*

Samantha MacBride, *Recycling Reconsidered: The Present Failure and Future Promise of Environmental Action in the United States*

Andrew Karvonen, *Politics of Urban Runoff: Nature, Technology, and the Sustainable City*

Daniel Schneider, *Hybrid Nature: Sewage Treatment and the Contradictions of the Industrial Ecosystem*

Catherine Tumber, *Small, Gritty, and Green: The Promise of America's Smaller Industrial Cities in a Low-Carbon World*

Sam Bass Warner and Andrew H. Whittemore, *American Urban Form: A Representative History*

John Pucher and Ralph Buehler, eds., *City Cycling*

Stephanie Foote and Elizabeth Mazzolini, eds., *Histories of the Dustheap: Waste, Material Cultures, Social Justice*

David J. Hess, *Good Green Jobs in a Global Economy: Making and Keeping New Industries in the United States*

Joseph F. C. DiMento and Clifford Ellis, *Changing Lanes: Visions and Histories of Urban Freeways*

Joanna Robinson, *Contested Water: The Struggle against Water Privatization in the United States and Canada*

William B. Meyer, *The Environmental Advantages of Cities: Countering Commonsense Antiurbanism*

Rebecca L. Henn and Andrew J. Hoffman, eds., *Constructing Green: The Social Structures of Sustainability*

Peggy F. Barlett and Geoffrey W. Chase, eds., *Sustainability in Higher Education: Stories and Strategies for Transformation*

Isabelle Anguelovski, *Neighborhood as Refuge: Community Reconstruction, Place Remaking, and Environmental Justice in the City*

Kelly Sims Gallagher, *The Globalization of Clean Energy Technology: Lessons from China*